高等学校Java课程系列教材

Java设计模式
——面向对象的设计思想
第2版·微课视频版

◎ 耿祥义 张跃平 主编

清华大学出版社
北京

内 容 简 介

本书面向有一定 Java 语言基础和一定编程经验的读者,重点探讨在 Java 程序设计中怎样使用重要的设计模式。本书共28章。前 6 章是学习设计模式的一些必要知识准备,也是 Java 语言的一些重要的概念和核心技术;第 7~27 章探讨、讲解 GoF 的《设计模式》一书中所给出的设计模式(除了代理模式和解释器模式);第 28 章为 MVC 模式。本书的编写目的是让读者不仅学习怎样在软件设计中使用好设计模式,而且深刻地理解面向对象的设计思想,以便更好地使用面向对象语言解决设计中的诸多问题。本书提供了 44 小节、总计 880 分钟的微课视频,对每个模式都进行了细致的讲解,非常有利于读者掌握本书的内容。

本书可作为计算机相关专业教材,也可作为软件项目管理人员、软件开发工程师等专业人员的参考用书。

本书封面贴有清华大学出版社防伪标签,无标签者不得销售。
版权所有,侵权必究。举报: 010-62782989,beiqinquan@tup.tsinghua.edu.cn。

图书在版编目(CIP)数据

Java 设计模式:面向对象的设计思想:微课视频版/耿祥义,张跃平主编. —2 版. —北京:清华大学出版社,2023.4(2024.7重印)
高等学校 Java 课程系列教材
ISBN 978-7-302-60951-3

Ⅰ. ①J… Ⅱ. ①耿… ②张… Ⅲ. ①JAVA 语言－程序设计－高等学校－教材 Ⅳ. ①TP312.8

中国版本图书馆 CIP 数据核字(2022)第 088998 号

策划编辑:魏江江
责任编辑:王冰飞
封面设计:刘　键
责任校对:时翠兰
责任印制:丛怀宇

出版发行:清华大学出版社
　　　　网　　址:https://www.tup.com.cn,https://www.wqxuetang.com
　　　　地　　址:北京清华大学学研大厦 A 座　　邮　编:100084
　　　　社 总 机:010-83470000　　邮　购:010-62786544
　　　　投稿与读者服务:010-62776969,c-service@tup.tsinghua.edu.cn
　　　　质量反馈:010-62772015,zhiliang@tup.tsinghua.edu.cn
　　　　课件下载:https://www.tup.com.cn,010-83470236
印 装 者:三河市天利华印刷装订有限公司
经　　销:全国新华书店
开　　本:185mm×260mm　　印　张:15.5　　字　数:396 千字
版　　次:2009 年 5 月第 1 版　2023 年 4 月第 2 版　　印　次:2024 年 7 月第 3 次印刷
印　　数:29501~31500
定　　价:49.80 元

产品编号:096557-01

前言

　　党的二十大报告指出：教育、科技、人才是全面建设社会主义现代化国家的基础性、战略性支撑。必须坚持科技是第一生产力、人才是第一资源、创新是第一动力,深入实施科教兴国战略、人才强国战略、创新驱动发展战略,开辟发展新领域新赛道,不断塑造发展新动能新优势。高等教育与经济社会发展紧密相连,对促进就业创业、助力经济社会发展、增进人民福祉具有重要意义。

　　党的二十大报告中指出：教育、科技、人才是全面建设社会主义现代化国家的基础性、战略性支撑。必须坚持科技是第一生产力、人才是第一资源、创新是第一动力,深入实施科教兴国战略、人才强国战略、创新驱动发展战略,这三大战略共同服务于创新型国家的建设。高等教育与经济社会发展紧密相连,对促进就业创业、助力经济社会发展、增进人民福祉具有重要意义。

　　设计模式是从许多优秀的软件系统中总结出的成功的、可复用的设计方案,已经被成功应用于许多系统的设计中。目前,面向对象程序设计已经成为软件设计开发领域的主流,而学习设计模式无疑非常有助于软件开发人员使用面向对象语言开发出易维护、易扩展、易复用的代码。本书面向有一定 Java 语言基础和一定编程经验的读者,重点探讨在 Java 程序设计中怎样使用 GoF 的《设计模式》一书中所给出的设计模式。本书的编写目的是让读者不仅学习怎样在软件设计过程中使用设计模式,而且深刻地理解面向对象的设计思想,以便更好地使用面向对象语言解决设计中的诸多问题。

　　本书共 28 章。前 6 章是学习设计模式的一些必要知识准备,也是 Java 语言的一些重要的概念和核心技术；第 7～27 章探讨、讲解 GoF 的《设计模式》一书中所给出的设计模式(除了代理模式和解释器模式)；第 28 章为 MVC 模式。第 7～28 章的每章都包含四部分重要的内容,第一部分为概述,通过易于理解的问题讲解使用模式的动机；第二部分是模式的结构与使用,用一个易于理解的示例讲解模式的结构；第三部分阐述模式的优点；第四部分结合实际问题,使用设计模式给出一个有一定难度和实用价值的示例。

　　本书特色：

　　1. 知识结构

　　对于每个模式,借助恰如其分的场景给出模式的定义、结构以及优点,非常有利于读者正确掌握该模式。在此基础上,给出该模式的一个有实用价值的综合应用,进一步巩固学习效果。

　　2. 模式示例

　　为了说明模式的核心实质,本书精心研究了针对每个模式的示例,以便让读者结合这样的示例更好地理解和使用模式。本书的全部示例由编者自行编写完成,并在JDK 14环境下编译通过(JDK版本不能低于JDK 8)。

3. 微课视频

作者为每个模式精心录制了微课视频,视频总时长880分钟,这些微课视频对读者学习本书内容有很大的帮助。

4. 教学资源

为便于教学,本书提供丰富的配套资源,包括教学大纲、教学课件、电子教案和程序源码。

资源下载提示

课件等资源:扫描封底的"课件下载"二维码,在公众号"书圈"下载。

素材(源码)等资源:扫描目录上方的二维码下载。

视频等资源:扫描封底的文泉云盘防盗码,再扫描书中相应章节的二维码,可以在线学习。

本书示例代码及相关内容仅供学习Java设计模式使用,不得以任何方式抄袭出版。

希望本书能对读者学习和使用设计模式有所帮助,并请读者批评指正。

编 者

2023年2月

目录

第 1 章　对象的基本结构

1.1　对象的引用和对象的变量 1
1.2　具有相同引用的对象 4
1.3　上转型对象 6

第 2 章　抽象类与接口

2.1　抽象类 8
2.2　接口 10
2.3　抽象类与接口的比较 14

第 3 章　组合

3.1　对象的组合 15
3.2　组合关系是弱耦合关系 17
3.3　基于组合的击鼓传花 18

第 4 章　面向对象的几个基本原则

4.1　面向抽象编程原则 21
4.2　"开-闭"原则 23
4.3　"多用组合，少用继承"原则 26
4.4　高内聚-低耦合原则 29

第 5 章　UML 类图简介

5.1　类 30
5.2　接口 30
5.3　泛化关系 31
5.4　关联关系 31
5.5　依赖关系 31

5.6 实现关系 ……………………………………………………………………… 32
5.7 注释 …………………………………………………………………………… 32

第 6 章　设计模式简介

6.1 什么是设计模式 …………………………………………………………… 33
6.2 设计模式的起源 …………………………………………………………… 33
6.3 GoF 之书 ……………………………………………………………………… 34
6.4 学习设计模式的重要性 …………………………………………………… 34
6.5 合理使用模式 ……………………………………………………………… 35
6.6 模式与框架 ………………………………………………………………… 36

第 7 章　策略模式

7.1 概述 …………………………………………………………………………… 37
7.2 模式的结构与使用 ………………………………………………………… 39
　　7.2.1 策略模式的 UML 类图 ……………………………………………… 40
　　7.2.2 结构的描述 …………………………………………………………… 40
　　7.2.3 模式的使用 …………………………………………………………… 43
7.3 策略模式的优点 …………………………………………………………… 44
7.4 应用举例——老鼠走迷宫 ………………………………………………… 44

第 8 章　责任链模式

8.1 概述 …………………………………………………………………………… 50
8.2 模式的结构与使用 ………………………………………………………… 51
　　8.2.1 责任链模式的 UML 类图 …………………………………………… 51
　　8.2.2 结构的描述 …………………………………………………………… 52
　　8.2.3 模式的使用 …………………………………………………………… 54
8.3 责任链模式的优点 ………………………………………………………… 55
8.4 应用举例——现金找零 …………………………………………………… 56

第 9 章　访问者模式

9.1 概述 …………………………………………………………………………… 59
9.2 模式的结构与使用 ………………………………………………………… 60
　　9.2.1 访问者模式的 UML 类图 …………………………………………… 60
　　9.2.2 结构的描述 …………………………………………………………… 61
　　9.2.3 模式的使用 …………………………………………………………… 62
9.3 访问者模式的优点 ………………………………………………………… 63

9.4　应用举例——答卷与批卷 ……………………………………………… 64

第 10 章　状态模式

10.1　概述 …………………………………………………………………………… 67
10.2　模式的结构与使用 ……………………………………………………………… 68
　　10.2.1　状态模式的 UML 类图 ……………………………………………… 68
　　10.2.2　结构的描述 …………………………………………………………… 68
　　10.2.3　模式的使用 …………………………………………………………… 71
10.3　状态模式的优点 ………………………………………………………………… 72
10.4　应用举例——咖啡自动售货机 ………………………………………………… 72

第 11 章　装饰模式

11.1　概述 …………………………………………………………………………… 76
11.2　模式的结构与使用 ……………………………………………………………… 77
　　11.2.1　装饰模式的 UML 类图 ……………………………………………… 77
　　11.2.2　结构的描述 …………………………………………………………… 78
　　11.2.3　模式的使用 …………………………………………………………… 79
11.3　装饰模式的优点 ………………………………………………………………… 79
11.4　应用举例——读取单词表 ……………………………………………………… 80

第 12 章　生成器模式

12.1　概述 …………………………………………………………………………… 84
12.2　模式的结构与使用 ……………………………………………………………… 85
　　12.2.1　生成器模式的 UML 类图 …………………………………………… 85
　　12.2.2　结构的描述 …………………………………………………………… 86
　　12.2.3　模式的使用 …………………………………………………………… 88
12.3　生成器模式的优点 ……………………………………………………………… 89
12.4　应用举例——日历牌 …………………………………………………………… 90

第 13 章　工厂方法模式

13.1　概述 …………………………………………………………………………… 96
13.2　模式的结构与使用 ……………………………………………………………… 97
　　13.2.1　工厂方法模式的 UML 类图 ………………………………………… 97
　　13.2.2　结构的描述 …………………………………………………………… 97
　　13.2.3　模式的使用 …………………………………………………………… 99
13.3　工厂方法模式的优点 …………………………………………………………… 99

13.4　应用举例——创建药品对象 …………………………………………………… 100

第 14 章　抽象工厂模式

14.1　概述 ……………………………………………………………………………… 103
14.2　模式的结构与使用 ……………………………………………………………… 104
　　14.2.1　抽象工厂模式的 UML 类图 ……………………………………………… 104
　　14.2.2　结构的描述 ……………………………………………………………… 105
　　14.2.3　模式的使用 ……………………………………………………………… 108
14.3　抽象工厂模式的优点 …………………………………………………………… 109
14.4　应用举例——存款凭证 ………………………………………………………… 109

第 15 章　命令模式

15.1　概述 ……………………………………………………………………………… 114
15.2　模式的结构与使用 ……………………………………………………………… 115
　　15.2.1　命令模式的 UML 类图 …………………………………………………… 115
　　15.2.2　结构的描述 ……………………………………………………………… 115
　　15.2.3　模式的使用 ……………………………………………………………… 118
15.3　命令模式的优点 ………………………………………………………………… 119
15.4　应用举例——开灯与关灯 ……………………………………………………… 119

第 16 章　桥接模式

16.1　概述 ……………………………………………………………………………… 123
16.2　模式的结构与使用 ……………………………………………………………… 124
　　16.2.1　桥接模式的 UML 类图 …………………………………………………… 124
　　16.2.2　结构的描述 ……………………………………………………………… 125
　　16.2.3　模式的使用 ……………………………………………………………… 126
16.3　桥接模式的优点 ………………………………………………………………… 127
16.4　应用举例——绘制简单图形 …………………………………………………… 128

第 17 章　单件模式

17.1　概述 ……………………………………………………………………………… 132
17.2　模式的结构与使用 ……………………………………………………………… 132
　　17.2.1　单件模式的 UML 类图 …………………………………………………… 133
　　17.2.2　结构的描述 ……………………………………………………………… 133
　　17.2.3　模式的使用 ……………………………………………………………… 134
17.3　单件模式的优点 ………………………………………………………………… 135

17.4　应用举例——多线程争冠军 135

第18章　适配器模式

18.1　概述 139
18.2　模式的结构与使用 139
　　18.2.1　适配器模式的 UML 类图 140
　　18.2.2　结构的描述 140
　　18.2.3　模式的使用 141
18.3　适配器模式的优点 142
18.4　应用举例——替换旧的加密、解密接口 142

第19章　模板方法模式

19.1　概述 145
19.2　模式的结构与使用 146
　　19.2.1　模板方法模式的 UML 类图 146
　　19.2.2　结构的描述 146
　　19.2.3　模式的使用 148
19.3　模板方法模式的优点 149
19.4　应用举例——数据挖掘 149

第20章　外观模式

20.1　概述 155
20.2　模式的结构与使用 156
　　20.2.1　外观模式的 UML 类图 156
　　20.2.2　结构的描述 156
　　20.2.3　模式的使用 158
20.3　外观模式的优点 159
20.4　应用举例——解析文件 159

第21章　中介者模式

21.1　概述 162
21.2　模式的结构与使用 163
　　21.2.1　中介者模式的 UML 类图 163
　　21.2.2　结构的描述 163
　　21.2.3　模式的使用 165
21.3　中介者模式的优点 165

21.4 应用举例——协调复制、剪切与粘贴 …………………………………………… 166

第 22 章 迭代器模式

22.1 概述 ………………………………………………………………………………… 169
22.2 模式的结构与使用 …………………………………………………………………… 170
  22.2.1 迭代器模式的 UML 类图 …………………………………………………… 170
  22.2.2 结构的描述 …………………………………………………………………… 170
  22.2.3 模式的使用 …………………………………………………………………… 173
22.3 迭代器模式的优点 …………………………………………………………………… 173
22.4 应用举例——使用多个集合存储对象 ……………………………………………… 173

第 23 章 组合模式

23.1 概述 ………………………………………………………………………………… 177
23.2 模式的结构与使用 …………………………………………………………………… 178
  23.2.1 组合模式的 UML 类图 ……………………………………………………… 178
  23.2.2 结构的描述 …………………………………………………………………… 178
  23.2.3 模式的使用 …………………………………………………………………… 180
23.3 组合模式的优点 ……………………………………………………………………… 182
23.4 应用举例——苹果树的苹果价值 …………………………………………………… 182

第 24 章 观察者模式

24.1 概述 ………………………………………………………………………………… 188
24.2 模式的结构与使用 …………………………………………………………………… 189
  24.2.1 观察者模式的 UML 类图 …………………………………………………… 189
  24.2.2 结构的描述 …………………………………………………………………… 189
  24.2.3 模式的使用 …………………………………………………………………… 192
24.3 观察者模式的优点 …………………………………………………………………… 192
24.4 应用举例——求面积服务中心 ……………………………………………………… 193

第 25 章 原型模式

25.1 概述 ………………………………………………………………………………… 197
25.2 模式的结构与使用 …………………………………………………………………… 198
  25.2.1 原型模式的 UML 类图 ……………………………………………………… 198
  25.2.2 结构的描述 …………………………………………………………………… 199
  25.2.3 模式的使用 …………………………………………………………………… 200
25.3 原型模式的优点 ……………………………………………………………………… 201

25.4　应用举例——克隆容器 201

第 26 章　备忘录模式

26.1　概述 204
26.2　模式的结构与使用 204
　　26.2.1　备忘录模式的 UML 类图 205
　　26.2.2　结构的描述 205
　　26.2.3　模式的使用 207
26.3　备忘录模式的优点 208
26.4　应用举例——使用备忘录实现 undo 操作 209

第 27 章　享元模式

27.1　概述 212
27.2　模式的结构与使用 213
　　27.2.1　享元模式的 UML 类图 213
　　27.2.2　结构的描述 213
　　27.2.3　模式的使用 215
27.3　享元模式的优点 216
27.4　应用举例——创建化合物 216

第 28 章　MVC 模式

28.1　概述 220
28.2　模式的结构与使用 220
　　28.2.1　MVC 模式的示意图 220
　　28.2.2　结构的描述 221
　　28.2.3　模式的使用 223
28.3　MVC 模式的优点 224
28.4　应用举例——老鼠走迷宫 224

参考文献 233

第 1 章 对象的基本结构

1.1 对象的引用和对象的变量

类封装了一类事物共有的属性和行为(相比 C 语言的结构体前进了一大步),并用一定的语法格式描述所封装的属性和行为。封装的关键是抓住事物的两个方面:属性和行为,即数据以及在数据上所进行的操作。

类是用于创建对象(对象也称类的实例)的一种数据类型(高级语言总是先有类型,再定义数据),也是 Java 语言中最重要的数据类型。类声明的变量称作对象变量,简称对象。

下面用简单的 CarSUV 类强调一下对象的基本结构。

```java
public class CarSUV {
    int speed;
    int weight = 200;
    int upSpeed(int n){
        if(n<=0||n>=260)
            return speed;
        speed += n;
        return speed>=260?260:speed;
    }
}
```

上述 CarSUV 类是用户定义的一种数据类型,下列代码用这个类型声明了两个对象,分别是 redCar 和 blueCar:

```java
CarSUV redCar = null;
CarSUV blueCar = null;
```

内存示意图如图 1.1 所示。

此时,系统认为 redCar 对象和 blueCar 对象中的数据是 null,称这样的对象是空对象。程序要避免使用空对象,即在使用对象之前,要确保为对象分配了变量(也称为对象分配实体)。下列代码为 redCar 对象和 blueCar 对象分配变量:

```java
redCar = new CarSUV();
blueCar = new CarSUV();
```

图 1.1 类声明的 redCar 对象和 blueCar 对象

new 是 Java 中的一个关键字,也是一个运算符,new 只能与构造方法进行运算,其作用非常类似 C 语言中的分配内存的函数 *calloc()(*calloc()分配内存,并初始化所分配的内存;*malloc()分配内存,但不初始化所分配的内存),即为 CarSUV 类中的成员变量 speed、weight 分配内存,为方法 speedUp() 分配入口地址。new 运算符为成员变量 speed、weight 分配内存后,进行初始化,然后执行构造方法,最后计算出一个称作"引用"的 int 型的值。程序需要将这个引用赋值给某个对象,以确保所分配的成员变量 speed、weight 是属于这个对象的,即确保 speed 和 weight 是分配给该对象的变量。CarSUV 类可以声明多个不同的对象,并使用 new 运算符为这些对象分配各自的变量,分配给不同对象的变量各自占有不同的内存空间(给对象分配变量也称创建对象)。此时的内存示意图如图 1.2 所示(图中画的小汽车是为了形象而画)。

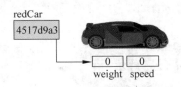
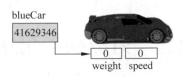

图 1.2 为 redCar 对象和 blueCar 对象分配变量

对象(变量)负责存放引用,以确保对象可以操作分配给该对象的变量(也称分配给对象的变量是对象的实体)以及调用类中的方法。图 1.2 中的箭头示意对象可以使用访问符"."访问分配给自己的变量。执行下列代码:

```
redCar.weight = 100;
blueCar.weight = 120;
```

那么 redCar 对象的 weight 的值是 100,blueCar 对象的 weight 的值是 120。此时的内存示意图如图 1.3 所示。

当对象调用方法时,方法中出现的成员变量就是指分配给该对象的变量(体现封装性)。执行下列代码:

```
redCar.upSpeed(60);
blueCar.upSpeed(80);
```

那么 redCar 对象的 speed 的值由 0 变成 60,blueCar 对象的 speed 的值从 0 变成 80。此时的内存示意图如图 1.4 所示。

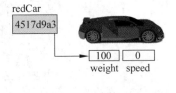
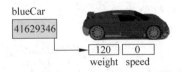

图 1.3 redCar 对象和 blueCar 对象各自访问自己的 weight 变量

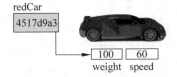
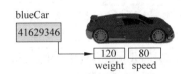

图 1.4 redCar 对象和 blueCar 对象调用 upSpeed() 方法

需要注意的是,当用下列代码输出对象中存放的引用时:

```
System.out.println(redCar);
```

得到的结果不完全是引用值,而是一个字符序列 CarSUV@4517d9a3,该字符序列给引用值(十六进制)添加了前缀信息"类名@"。其原因是:

```
System.out.println(redCar);
```

等价于:

```
System.out.println(redCar.toString());
```

public String toString()方法是 Object 类提供的一个方法,CarSUV 类可以重写该方法,使得 toString()方法只返回引用值的字符序列表示。例如在 CarSUV 类中,进行如下的重写:

```
public String toString() {                       //重写 Object 类的方法
    String str = super.toString();
    String backStr = str.substring(str.indexOf("@") + 1);
    return backStr;
}
```

如果需要按整数输出对象的引用(一个整型值,有特殊意义的整数),可以使用 System 类提供的静态方法 identityHashCode()。例如:

```
int addr = System.identityHashCode(redCar);
```

返回的 addr 是 int 型值。

要很好地理解对象的基本结构,即对象(变量)负责存放引用,以确保对象可以操作分配给该对象的变量以及调用类中的方法。不要把对象和分配给对象的变量混淆(分配给对象的变量仅仅是对象的一部分)。避免使用匿名对象,例如,代码

```
new Car().weight = 100;
new Car().weight = 120;
```

的效果是,两个不同的对象(匿名对象)分别设置自己的 weight,而代码

```
redCar.weight = 100;
redCar.weight = 120;
```

的效果是,redCar 对象首先将自己的 weight 设置为 100,然后又重新设置为 120。

注意:在学习面向对象语言的过程中,一个简单的理念是需要完成某种任务时,首先要想到谁去完成任务,即哪个对象去完成任务;提到数据,首先要想到这个数据是哪个对象的。

例 1-1 演示了对象的基本结构。程序运行效果如图 1.5 所示。

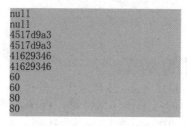

图 1.5 程序运行效果

例 1-1
CarSUV.java

```java
public class CarSUV {
    int speed;
    int weight = 200;
    int upSpeed(int n){
        if(n <= 0 || n >= 260) {
            return speed;
        }
        speed += n;
        return speed >= 260?260:speed;
    }
    public String toString() {        //重写 Object 类的方法
        String str = super.toString();
        String backStr = str.substring(str.indexOf("@") + 1);
        return backStr;
    }
}
```

Example1_1.java

```java
public class Example1_1 {
    public static void main(String args[]) {
        CarSUV redCar = null;
        CarSUV blueCar = null;
        System.out.println(redCar);
        System.out.println(blueCar);
        redCar = new CarSUV();
        blueCar = new CarSUV();
        System.out.println(redCar);
        int addr = System.identityHashCode(redCar);
        System.out.printf(" %x\n",addr);
        System.out.println(blueCar);
        addr = System.identityHashCode(blueCar);
        System.out.printf(" %x\n",addr);
        redCar.weight = 100;
        blueCar.weight = 120;
        int m = redCar.upSpeed(60);
        int n = blueCar.upSpeed(80);
        System.out.println(m);
        System.out.println(redCar.speed);
        System.out.println(n);
        System.out.println(blueCar.speed);
    }
}
```

1.2 具有相同引用的对象

没有实体(没有被分配变量)的对象称作空对象。程序要避免让一个空对象去调用方法产生行为,否则在运行时会出现 NullPointerException 异常。由于对象可以被动态地分配变量

(实体),所以 Java 编译器对空对象不做检查(不会出现编译错误)。因此,在编写程序时要避免使用空对象。

一个类声明的两个对象如果具有相同的引用,二者就具有完全相同的变量(实体)。当程序用一个类为两个对象 object1 和 object2 分配变量后,二者的引用是不同的,如图 1.6 所示。

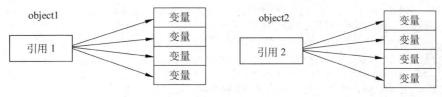

图 1.6　具有不同引用的对象

在 Java 中,对于同一个类的两个对象 object1 和 object2,允许进行如下的赋值操作:

object2 = object1;

这样,object2 中存放的是 object1 的值,即 object1 对象的引用,因此,object2 所拥有的变量(实体)就和 object1 完全一样了,如图 1.7 所示。

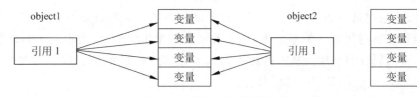

图 1.7　具有相同引用的对象

Java 有所谓的"垃圾收集"机制,这种机制周期地检测某个实体是否已不再被任何对象所拥有(引用),如果发现这样的实体,就释放实体占有的内存。

例 1-2 演示了如果类声明的两个对象具有相同的引用,二者就具有完全相同的变量。程序运行效果如图 1.8 所示。

```
p1的引用:Point@3a71f4dd
p2的引用:Point@7adf9f5f
p1的x,y坐标:1111,2222
p2的x,y坐标:-100,-200
将p2的引用赋给p1后:
p1的引用:7adf9f5f
p2的引用:7adf9f5f
p1的x,y坐标:-100,-200
p2的x,y坐标:-100,-200
```

图 1.8　程序运行效果

例 1-2

Example1_2.java

```java
class Point {
    int x,y;
    void setXY(int m,int n){
        x = m;
        y = n;
    }
}
public class Example1_2 {
    public static void main(String args[]) {
        Point p1 = null,p2 = null;
        p1 = new Point();
        p2 = new Point();
        System.out.println("p1 的引用:" + p1);
```

```
        System.out.println("p2 的引用:" + p2);
        p1.setXY(1111,2222);
        p2.setXY(-100,-200);
        System.out.println("p1 的 x,y 坐标:" + p1.x + "," + p1.y);
        System.out.println("p2 的 x,y 坐标:" + p2.x + "," + p2.y);
        p1 = p2;                    //使得对象 p1 和 p2 的引用相同
        System.out.println("将 p2 的引用赋给 p1 后: ");
        int address = System.identityHashCode(p1);
        System.out.printf("p1 的引用:%x\n",address);
        address = System.identityHashCode(p2);
        System.out.printf("p2 的引用:%x\n",address);
        System.out.println("p1 的 x,y 坐标:" + p1.x + "," + p1.y);
        System.out.println("p2 的 x,y 坐标:" + p2.x + "," + p2.y);
    }
}
```

1.3 上转型对象

我们经常说"老虎是动物""狗是动物"等。若动物类是老虎类的父类,这样说当然正确,因为人们习惯地称子类与父类的关系是"is-a"关系。但需要注意的是,当说老虎是动物时,老虎将失掉老虎独有的属性和功能。从人的思维方式来看,说"老虎是动物"属于上溯思维方式,这种思维方式类似 Java 语言中的上转型对象。

假设 Animal 类是 Tiger 类的父类(或间接父类),当用子类创建一个对象,并把这个对象的引用赋值到父类声明的对象中时,例如:

```
Animal animal;
animal = new Tiger();
```

或

```
Animal animal;
Tiger tiger = new Tiger();
animal = tiger;
```

称 animal 是对象 tiger 的上转型对象(好比说"老虎是动物")。

对象的上转型对象的实体是子类负责创建的,但上转型对象会失去原对象的一些属性和功能(上转型对象相当于子类对象的一个"简化"对象)。上转型对象操作子类继承的方法或子类重写的实例方法,其作用等价于子类对象去调用这些方法。因此,如果子类重写了父类的某个实例方法,当对象的上转型对象调用这个实例方法时,一定是调用了子类重写的实例方法。

上转型对象的实体本质上是子类分配的实体,只是少了一部分实体而已,因此,对于

```
Animal animal;
animal = new Tiger();
```

下列表达式的值是 true:

```
animal instanceof Tiger
```

注意：如果子类重写了父类的静态方法，那么子类对象的上转型对象不能调用子类重写的静态方法，只能调用父类的静态方法。

例 1-3 中使用了上转型对象。程序运行效果如图 1.9 所示。

```
这个Dog@31befd9f是狗吗?true
wang wang.....
这个Dog@31befd9f是猫吗?false
这个Cat@1fb3ebeb是猫吗?true
miao miao.....
```

图 1.9　程序运行效果

例 1-3
Example1_3.java

```java
class Animal {
    void cry() {
        System.out.println("动物的叫声是怎样的?");
    }
}
class Dog extends Animal {
    void cry() {
        System.out.println("wang wang.....");
    }
}
class Cat extends Animal {
    void cry() {
        System.out.println("miao miao.....");
    }
}
public class Example1_3 {
    public static void main(String args[]) {
        Animal animal;
        animal = new Dog();
        boolean isOk = animal instanceof Dog;
        System.out.println("这个" + animal + "是狗吗?" + isOk);
        animal.cry();
        isOk = animal instanceof Cat;
        System.out.println("这个" + animal + "是猫吗?" + isOk);
        animal = new Cat();
        isOk = animal instanceof Cat;
        System.out.println("这个" + animal + "是猫吗?" + isOk);
        animal.cry();
    }
}
```

第 2 章　抽象类与接口

2.1　抽象类

用关键字 abstract 修饰的类称为抽象类（abstract 类），例如：

```
abstract class A {
   ...
}
```

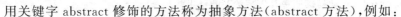

用关键字 abstract 修饰的方法称为抽象方法（abstract 方法），例如：

```
abstract int min(int x, int y);
```

对于抽象方法，只允许声明，不允许实现（没有方法体），而且不允许使用 final 和 abstract 同时修饰一个方法或类，也不允许使用 static 和 private 修饰 abstract 方法，即抽象方法必须是非 private 的实例方法（访问权限必须高于 private）。

1. 抽象类中的抽象方法

和普通类（非抽象类）相比，抽象类中可以有抽象方法（非抽象类中不可以有抽象方法），也可以有非抽象方法。

下面的 A 类中的 min() 方法是抽象方法，max() 方法是普通方法（非抽象方法）。

```
abstract class A {
   abstract int min(int x, int y);
   int max(int x, int y) {
      return x > y?x:y;
   }
}
```

注意：抽象类中也可以没有抽象方法。

2. 抽象类不能用 new 运算符创建对象

对于抽象类，不能使用 new 运算符创建该类的对象。如果一个非抽象类是某个抽象类的子类，那么它必须重写父类的抽象方法，给出方法体，这就是为什么不允许使用 final 和 abstract 同时修饰一个方法或类的原因。

3. 抽象类的子类

如果一个非抽象类是抽象类的子类，它必须重写父类的抽象方法，即去掉抽象方法的 abstract 修饰，并给出方法体。如果一个抽象类是抽象类的子类，它可以重写父类的抽象方法，也可以继承父类的抽象方法。

4. 抽象类与上转型对象

尽管抽象类不能使用 new 运算符创建对象，但抽象类声明的对象可以作为子类对象的上转型对象,那么该对象就可以调用子类重写的方法。

5. 理解抽象类

理解抽象类的意义非常重要。理解抽象类的关键点是:

(1) 抽象类可以抽象出重要的行为标准,该行为标准用抽象方法来表示。即抽象类封装了子类必须有的行为标准。

(2) 抽象类声明的对象可以成为其子类的对象的上转型对象,调用子类重写的方法,即体现子类根据抽象类中的行为标准给出的具体行为。

人们已经习惯给别人介绍数量标准,例如,在介绍人的时候,可以说,人的身高是多少,体重是多少,但是学习了类以后,也要习惯介绍行为标准。所谓行为标准,仅仅是方法的名字、方法的类型而已。例如,人具有 run()行为,或 speak()行为,但仅仅说出行为标准,不要说出 speak()行为的具体体现,即不要说 speak()行为是用英语说话还是用中文说话,这样的行为标准就是抽象方法(没有方法体的方法)。这样一来,开发者可以把主要精力放在一个应用中需要哪些行为标准(不用关心行为的细节),不仅节省时间,而且非常有利于设计出易维护、易扩展的程序(见后面的设计模式相关的章节,例如第 7 章,策略模式)。抽象类中的抽象方法可以由子类实现,即行为标准的实现由子类完成。

一个男孩要找女朋友,他可以提出一些行为标准,例如,女朋友具有 speak()和 cooking()行为,但可以不给出 speak()和 cooking()行为的细节。例 2-1 使用抽象类封装了男孩对女朋友的行为要求,即封装了他要找的任何具体女朋友都应该具有的行为。程序运行效果如图 2.1 所示。

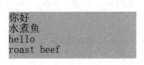

图 2.1 程序运行效果

例 2-1

Example2_1.java

```java
abstract class GirlFriend {                              //抽象类,封装了两个行为标准
    abstract void speak();
    abstract void cooking();
}
class ChinaGirlFriend extends GirlFriend {
    void speak(){
        System.out.println("你好");
    }
    void cooking(){
        System.out.println("水煮鱼");
    }
}
class AmericanGirlFriend extends GirlFriend {
    void speak(){
        System.out.println("hello");
    }
    void cooking(){
        System.out.println("roast beef");
    }
}
class Boy {
    GirlFriend friend;
```

```
        void setGirlfriend(GirlFriend f){
            friend = f;
        }
        void showGirlFriend() {
            friend.speak();
            friend.cooking();
        }
    }
    public class Example2_1 {
        public static void main(String args[]) {
            GirlFriend girl = new ChinaGirlFriend();          //girl 是上转型对象
            Boy boy = new Boy();
            boy.setGirlfriend(girl);
            boy.showGirlFriend();
            girl = new AmericanGirlFriend();                  //girl 是上转型对象
            boy.setGirlfriend(girl);
            boy.showGirlFriend();
        }
    }
```

2.2 接口

使用关键字 interface 定义一个接口。接口的定义和类的定义很相似,分为接口声明和接口体,例如:

```
interface Com {
    ...
}
```

1. 接口声明

定义接口包含接口声明和接口体,和类不同的是,定义接口时使用关键字 interface 来声明自己是一个接口,格式如下:

```
interface 接口的名字
```

2. 接口体

1) 接口体中的抽象方法和常量

接口中可以有抽象方法(在 JDK 8 版本之前,接口体中只可以有抽象方法),接口中所有的抽象方法的访问权限一定都是 public,而且允许省略抽象方法的 public 和 abstract 修饰符。接口体中所有的 static 常量的访问权限一定都是 public,而且允许省略 public、final 和 static 修饰符,因此,接口中不会有变量。例如:

```
interface Com {
    public static final int MAX = 100;                //等价写法: int MAX = 100;
    public abstract void add();                       //等价写法: void add();
    public abstract float sum(float x ,float y);
     //等价写法: float sum(float x ,float y);
}
```

2) 接口体中的 default 实例方法

从 JDK 8 版本开始,允许使用 default 关键字、在接口体中定义称作 default 的实例方法(不可以定义 default 的 static 方法)。和普通的实例方法相比,default 的实例方法是用关键字 default 修饰的带方法体的实例方法。default 实例方法的访问权限必须是 public(允许省略 public 修饰符)。例如,下列接口中的 max 方法就是 default 实例方法:

```java
interface Com {
    public final int MAX = 100;
    public abstract void add();
    public abstract float sum(float x ,float y);
    public default int max(int a, int b) {          //default 方法
        return a > b?a:b;
    }
}
```

注意:不可以省略 default 关键字,因为接口里不允许定义通常的带方法体的 public 实例方法。

3) 接口体中的静态(static)方法

从 JDK 8 版本开始,允许在接口体中定义静态方法。例如,下列接口中的 f() 方法就是静态方法:

```java
public interface Com {
    public static final int MAX = 100;
    public abstract void on();
    public abstract float sum(float x ,float y);
    public default int max(int a, int b) {
        return a > b?a:b;
    }
    public static void f() {                //static 方法
        System.out.println("注意是从 JDK 8 开始的");
    }
}
```

4) 接口体中的私有(private)方法

从 JDK 9 版本开始,允许在接口体中定义私有的方法,其目的是配合接口中的 default 的实例方法,即接口可以将某些算法封装在私有的方法中,供接口中的 default 的实例方法调用,实现算法的复用。

3. 实现接口

在 Java 语言中,接口由类实现以便使用接口中的方法。一个类需要在类声明中使用关键字 implements 声明该类实现一个或多个接口。如果实现多个接口,用逗号隔开接口名。例如,A 类实现 Com 和 Addable 两个接口:

```java
class A implements Com,Addable
```

再如,Animal 的 Dog 子类实现 Eatable 和 Sleepable 接口:

```java
class Dog extends Animal implements Eatable,Sleepable
```

如果一个类实现了某个接口，那么这个类就自然拥有了接口中的常量和 default 关键字修饰的方法(但去掉了 default 关键字)，该类也可以重写接口中 default 修饰的方法(注意，重写时需要去掉 default 关键字)。如果一个非抽象类实现了某个接口，那么这个类必须重写该接口的所有抽象方法，即去掉修饰方法的 abstract 关键字，给出方法体。如果一个抽象类实现了某个接口，该类可以选择重写接口的抽象方法或直接用接口的抽象方法。

需要特别注意的是，类实现接口，但类并不拥有接口的静态方法和私有方法。接口中除了私有方法，其他方法的访问权限默认都是 public 的，重写时不可省略 public(否则就降低了访问权限，这是不允许的)。

实现接口的非抽象类一定要重写接口的抽象方法，因此也称这个类实现了接口。

4. 接口回调

和类一样，接口也是 Java 中一种重要的数据类型，用接口声明的变量称作接口变量。那么，接口变量中可以存放怎样的数据呢？

接口属于引用型变量，接口变量中可以存放实现该接口的类的实例的引用，即存放对象的引用。例如，假设 Com 是一个接口，那么就可以用 Com 声明一个变量：

```
Com com;
```

其内存模型如图 2.2 所示，称此时的 com 是一个空接口，因为 com 变量中还没有存放实现该接口的类的实例(对象)的引用。

假设 ImpleCom 类是实现 Com 接口的类，用 ImpleCom 创建名字为 object 的对象：

```
ImpleCom object = new ImpleCom();
```

那么 object 对象不仅可以调用 ImpleCom 类中原有的方法，而且可以调用 ImpleCom 类实现的接口方法，如图 2.3 所示。

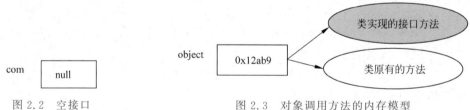

图 2.2　空接口　　　　　　　图 2.3　对象调用方法的内存模型

"接口回调"一词是借用了 C 语言中指针回调的术语，表示一个变量的地址在某一个时刻存放在一个指针变量中，那么指针变量就可以间接操作该变量中存放的数据。

在 Java 语言中，接口回调是指可以把实现某一接口的类创建的对象的引用赋值给该接口声明的接口变量，那么该接口变量就可以调用被类实现的接口方法以及接口提供的 default 方法或类重写的 default 方法。实际上，当接口变量调用被类实现的接口方法时，就是通知相应的对象调用这个方法。

例如，将上述 object 对象的引用赋值给 com 接口：

```
com = object;
```

那么内存模型如图 2.4 所示，箭头示意接口 com 变量可以调用类实现的接口方法(这一过程称为接口回调）。但是，接口无法调用类中的其他的非接口方法。

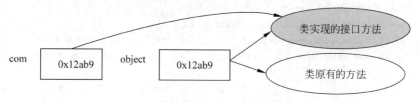

图 2.4　接口回调的内存模型

接口回调非常类似于上转型对象调用子类重写的方法（见 1.3 节）。

注意：如果接口 com 中存放了实现该接口的 ImpleCom 类的对象的引用，那么表达式 com instanceof ImpleCom 的值是 true。

例 2-2 使用了接口的回调技术，程序运行效果如图 2.5 所示。

```
true
11.23和22.78的算术平均值:17.01
false
true
11.23和22.78的几何平均值:15.99
```

图 2.5　程序运行效果

例 2-2

Example2_2.java

```java
interface ComputerAverage {
    public double average(double a, double b);
}
class A implements ComputerAverage {
    public double average(double a, double b) {
        double aver = 0;
        aver = (a+b)/2;
        return aver;
    }
}
class B implements ComputerAverage {
    public double average(double a, double b) {
        double aver = 0;
        aver = Math.sqrt(a * b);
        return aver;
    }
}
public class Example2_2 {
    public static void main(String args[]) {
        ComputerAverage computer;
        double a = 11.23,b = 22.78;
        computer = new A();
        System.out.println(computer instanceof A);
        double result = computer.average(a,b);              //接口回调 average()方法
        System.out.printf(" %5.2f 和 %5.2f 的算术平均值:%5.2f\n",a,b,result);
```

```
        computer = new B();
        System.out.println(computer instanceof A);
        System.out.println(computer instanceof B);
        result = computer.average(a,b);              //接口回调 average()方法
        System.out.printf(" %5.2f 和 %5.2f 的几何平均值:%5.2f",a,b,result);
    }
}
```

2.3 抽象类与接口的比较

　　抽象类和接口都可以有抽象方法。接口中只可以有常量,不可以有变量;而抽象类中既可以有常量,也可以有变量。抽象类中可以有非抽象的,且不能用 default 关键字修饰的实例方法;而接口中不可以有非抽象的,不用 default 关键字修饰的 public 实例方法。

　　在设计程序时应当根据具体的分析确定是使用抽象类还是接口。抽象类除了提供重要的需要子类重写的抽象方法外,还提供子类可以继承的变量和非抽象方法。如果某个问题需要使用继承才能更好地解决,例如,子类除了需要重写父类的抽象方法,还需要从父类继承一些变量或一些重要的非抽象方法,就可以考虑用抽象类;如果某个问题不需要继承,只是需要若干个类给出某些重要的抽象方法的实现细节,就可以考虑使用接口。

第 3 章 组合

在实际生活中经常能见到对象组合的例子,例如一个公司有若干职员,一个房屋有若干家具等。组合是面向对象中的一个重要手段,通过组合,可以让对象之间进行必要的交互。

3.1 对象的组合

类的成员变量可以是 Java 允许的任何数据类型,因此,一个类可以把对象作为自己的成员变量。如果用这样的类创建对象,那么该对象中就会有其他对象,也就是说,该对象将其他对象作为自己的组成部分(这就是人们常说的Has-A)。一个对象 a 通过组合对象 b 来复用对象 b 的方法,即对象 a 委托对象 b 调用其方法。当前对象随时可以更换所组合的对象,使得当前对象与所组合的对象是弱耦合关系。

现在,对圆锥体做一个抽象:

- 属性:底圆,高
- 操作:计算体积

那么,圆锥体的底圆应当是一个对象,例如 Circle 类声明的对象;圆锥体的高可以是 double 型的变量,即圆锥体将 Circle 类的对象作为自己的成员。

例 3-1 中,Circular 类负责创建"圆锥体"对象;Application.java 是主类。在主类的 main 方法中使用 Circle 类创建一个"圆"对象 circle,使用 Circular 类创建一个"圆锥"对象 circular,然后 circular 调用 setBottom(Circle c)方法将 circle 的引用传递给 circular 的成员变量 bottom,即让 circular 组合 circle。程序运行效果如图 3.1 所示。

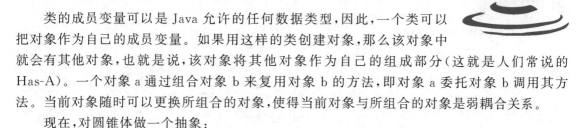

图 3.1 程序运行效果

例 3-1

Circle. java

```
public class Circle {
    double radius;
    double getArea() {
        double area = 3.14 * radius * radius;
        return area;
    }
}
```

Circular. java

```
public class Circular {                    //圆锥类
    Circle bottom;
    double height;
```

```java
    void setBottom(Circle c) {
        bottom = c;
    }
    void setHeight(double h) {
        height = h;
    }
    double getVolume() {
        return bottom.getArea() * height/3.0;
    }
}
```

Application.java

```java
public class Application {                           //主类
    public static void main(String args[]) {
        Circle circle = new Circle();
        circle.radius = 100;
        Circular circular = new Circular();
        circular.setBottom(circle);                  //将 circle 的引用传递给 circular 的 bottom
        circular.setHeight(6.66);
        System.out.printf("圆锥的体积:%5.3f\n",circular.getVolume());
    }
}
```

在上述 Application 类中,当执行代码

```java
Circle circle = new Circle(100);
```

后,内存中诞生了一个 circle(圆)对象,circle 的 radius(半径)是 100。内存中对象的模型如图 3.2 所示。

当执行代码

```java
Circular circular = new Circular(circle,6.66);
```

后,内存中又诞生了一个 circular 对象(圆锥),然后执行代码

```java
circular.setBottom(circle);
```

将 circle 对象的引用传递给 circular 对象的 bottom(底),因此,bottom 对象和 circle 对象就有同样的实体(radius)。内存中对象的模型如图 3.3 所示。

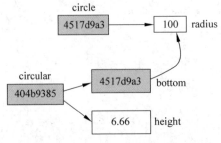

图 3.3 circular(圆锥)对象

3.2 组合关系是弱耦合关系

子类与父类的关系是强耦合关系,父子关系体现在类层次上,而不是对象层次上。组合关系最终体现在对象层次上,即一个对象将另一个对象作为自己的一部分。面向对象将对象与对象之间的组合关系归为弱耦合关系的主要原因有以下两点:

- 如果修改当前对象所组合的对象的类的代码,不必修改当前对象的类的代码。
- 当前对象可以在运行时指定所包含的对象。

公司和职员的关系是组合关系,一个公司可以根据公司的需求聘用或解聘某个职员。例如,Employee 是描述职员的类,Corp 是描述公司的类,让 Corp 和 Employee 类形成组合关系。那么,Corp 类的对象可以在运行时指定所包含的 Employee 子类的对象,即 Corp 类的实例可以调用 setEmployee(Employee em)方法在它的 employee 变量中存放任何 Employee 子类的对象的引用。具体代码见例 3-2,运行效果如图 3.4 所示。

```
本公司当前聘用的职员信息:
男性职员:周.
男性职员:吴.
女性职员:杨.
本公司当前聘用的职员信息:
男性职员:赵.
女性职员:孙.
```

图 3.4 程序运行效果

例 3-2

Employee. java

```java
public abstract class Employee{
    String name;
    public abstract void showMess();
}
```

MaleEmployee. java

```java
public class MaleEmployee extends Employee{
    public MaleEmployee(String name) {
        this.name = name;
    }
    public void showMess() {
        System.out.println("男性职员:" + name + ".");
    }
}
```

FemaleEmployee. java

```java
public class FemaleEmployee extends Employee {
    public FemaleEmployee(String name) {
        this.name = name;
    }
    public void showMess() {
        System.out.println("女性职员:" + name + ".");
    }
}
```

Corp. java

```java
public class Corp {
    Employee employee[];
```

```java
    public void setEmployee(Employee ...em) {
        employee = new Employee[em.length];
        for(int i = 0;i < em.length;i++)
            employee[i] = em[i];
    }
    public void outEmployeeMess() {
        System.out.println("本公司当前聘用的职员信息:");
        for(int i = 0;i < employee.length;i++)
            employee[i].showMess();
    }
}
```

Application.java

```java
public class Application {
  public static void main(String args[]){
     Corp corpSun = new Corp();
     corpSun.setEmployee
            (new MaleEmployee("周"),new MaleEmployee("吴"),new FemaleEmployee("杨"));
     corpSun.outEmployeeMess();
     corpSun.setEmployee
            (new MaleEmployee("赵"),new FemaleEmployee("孙"));
     corpSun.outEmployeeMess();
  }
}
```

3.3 基于组合的击鼓传花

击鼓传花是很多人玩过的游戏。一些人(不少于 2 人)围成圆圈,其中一人拿花,圈外一人背对着大家敲鼓。鼓响时众人开始依次传花,至鼓声停止为止。此时花在谁手中,谁就从圈中离开(实际娱乐中,常常要求出圈人上台表演节目);然后鼓声再次响起,圈中剩余的人继续依次传花。击鼓传花游戏如图 3.5 所示。

参与击鼓传花的人必须组合他的下一位,以便把花传给下一位;同时也必须组合上一位,以便自己退出时通知上一位,他的下一位发生了变化,见例 3-3 中的 Person 类。Flower 类的对象负责在 0,1,2,3,4,5,6,7 中产生一个随机数,当花在某人手中,而此时 Flower 对象产生的随机数刚好是 6(模拟鼓声停止),此人从圈中退出。程序运行效果如图 3.6 所示。

图 3.5 击鼓传花游戏

图 3.6 程序运行效果

例 3-3

Person.java

```java
public class Person {
    int personNumber;                              //代号
    Flower flowerInhand;                           //手中拿的花
    Person previousPerson;                         //组合传我花的人
    Person nextPerson;                             //组合接我花的人
    public void setFlower(Flower flower){
        flowerInhand = flower;
        int n = flowerInhand.getRamdomNumber();
        if(n == 6) {
            previousPerson.setNextPerson(nextPerson);
            nextPerson.setPreviousPerson(previousPerson);
            try{ Thread.sleep(1000);
            }
            catch(Exception exp){}
            System.out.println("代号" + personNumber + "的人退出击鼓传花");

        }
        if(this == nextPerson){
            System.out.println("代号" + personNumber + "是最后一个人");
            return;
        }
        nextPerson.setFlower(flowerInhand );        //传花给下一位
    }
    public void setPreviousPerson(Person person){
        previousPerson = person;
    }
    public void setNextPerson(Person person){
        nextPerson = person;
    }
}
```

Flower.java

```java
import java.util.Random;                            //负责产生随机数的 Random 类
public class Flower {
    Random random;                                  //负责产生随机数
    Flower(){
        random = new Random();
    }
    public int getRamdomNumber(){
        int number = random.nextInt(8);             //返回 0 到 7 中的某个数
        return number + 1;
    }
}
```

Application.java

```java
public class Application {
    public static void main(String args[]) {
```

```java
        Flower roseFlower = new Flower();
        int length = 12;
        if(length < 2)
            return;
        Person [] personInCircle = new Person[length];
        for(int i = 0;i < personInCircle.length;i++){
            personInCircle[i] = new Person();
            personInCircle[i].personNumber = i + 1;
        }
        for(int i = 0;i < personInCircle.length;i++){
            if(i == 0){
                personInCircle[i].setNextPerson(personInCircle[i + 1]);
                personInCircle[i].setPreviousPerson(personInCircle[length - 1]);
            }
            else if(i == length - 1){
                personInCircle[i].setNextPerson(personInCircle[0]);
                personInCircle[i].setPreviousPerson(personInCircle[i - 1]);
            }
            else {
                personInCircle[i].setNextPerson(personInCircle[i + 1]);
                personInCircle[i].setPreviousPerson(personInCircle[i - 1]);
            }
        }
        personInCircle[0].setFlower(roseFlower);
    }
}
```

第 4 章　面向对象的几个基本原则

本章给出面向对象设计的几个基本原则,了解这些基本原则有助于读者学会使用面向对象语言编写出易维护、易扩展和易复用的程序代码。许多模式都体现了本章所述的一些基本原则,但需要强调的一点是,本章介绍的这些原则是在许多设计中总结出的指导性原则,并不是任何设计都必须遵守的"语法"规定。

4.1　面向抽象编程原则

所谓面向抽象编程,是指当设计一个类时,不让该类面向具体的类,而是面向抽象类或接口,即所设计类中的重要数据是抽象类或接口声明的变量,而不是具体类声明的变量。

下面通过一个简单的问题说明面向抽象编程的思想。

已知一个 Circle 类,该类创建的对象 circle 调用 getArea()方法可以计算圆的面积。Circle 类的代码如下:

Circle.java

```java
public class Circle {
    double r;
    Circle(double r){
        this.r = r;
    }
    public double getArea() {
        return(3.14 * r * r);
    }
}
```

现在要设计一个 Pillar 类(柱类),该类的对象调用 getVolume()方法可以计算柱体的体积。Pillar 类的代码如下:

Pillar.java

```java
public class Pillar {
    Circle bottom;              //将 Circle 对象作为成员,bottom 是用具体类 Circle 声明的变量
    double height;
    Pillar (Circle bottom,double height) {
        this.bottom = bottom;
        this.height = height;
    }
    public double getVolume() {
        return bottom.getArea() * height;
    }
}
```

上述 Pillar 类中，bottom 是用具体类 Circle 声明的变量，如果不涉及用户需求的变化，上面 Pillar 类的设计没有什么不妥。但是在某个时候，用户希望 Pillar 类创建出底是三角形的柱体，显然上述 Pillar 类无法创建出这样的柱体，即上述 Pillar 类不能应对用户的这种需求。

现在重新设计 Pillar 类。首先，注意到柱体计算体积的关键是计算出底面积，一个柱体在计算底面积时不应该关心它的底是什么形状的具体图形，只应该关心这种图形是否具有计算面积的方法。因此，在设计 Pillar 类时，不应当让它的底是某个具体类声明的变量，一旦这样做，Pillar 类就依赖该具体类，缺乏弹性，难以应对需求的变化。

例 4-1 面向抽象重新设计 Pillar 类。首先编写一个抽象类 Geometry（或接口），该抽象类（接口）中定义了一个抽象的 getArea()方法。程序运行效果如图 4.1 所示。

矩形底的柱体的体积15312.0
圆形底的柱体的体积18212.0

图 4.1　程序运行效果

例 4-1

Geometry.java

```
public abstract class Geometry {           //如果使用接口,需用 interface 定义 Geometry
    public abstract double getArea();
}
```

现在 Pillar 类的设计者可以面向 Geometry 类编写代码，即 Pillar 类应当把 Geometry 对象作为自己的成员，该成员可以调用 Geometry 的子类重写的 getArea()方法（如果 Geometry 是一个接口，那么该成员可以回调实现 Geometry 接口的类所实现的 getArea()方法）。这样一来，Pillar 类就可以将计算底面积的任务指派给 Geometry 类的子类的实例（如果 Geometry 是一个接口，Pillar 类就可以将计算底面积的任务指派给实现 Geometry 接口的类的实例）。

以下 Pillar 类的设计不再依赖具体类，而是面向 Geometry 类，即 Pillar 类中的 bottom 是用抽象类 Geometry 声明的变量，而不是具体类声明的变量。重新设计的 Pillar 类的代码如下：

Pillar.java

```
public class Pillar {
    Geometry bottom;                       //bottom 是抽象类 Geometry 声明的变量
    double height;
    Pillar(Geometry bottom,double height) {
        this.bottom = bottom;
        this.height = height;
    }
    public double getVolume() {
        if(bottom!= null)
            return bottom.getArea() * height;
        else
            return Double.NaN;
    }
}
```

下列 Circle 类和 Rectangle 类都是 Geometry 的子类，二者都必须重写 Geometry 类的 getArea()方法计算各自的面积。

Circle.java

```
public class Circle extends Geometry {
    double r;
```

```java
    Circle(double r) {
        this.r = r;
    }
    public double getArea() {
        return(3.14 * r * r);
    }
}
```

Rectangle. java

```java
public class Rectangle extends Geometry {
    double a,b;
    Rectangle(double a,double b) {
        this.a = a;
        this.b = b;
    }
    public double getArea() {
        return a * b;
    }
}
```

现在就可以用 Pillar 类创建出具有矩形底或圆形底的柱体了,如下列代码所示。

Application. java

```java
public class Application{
    public static void main(String args[]){
        Pillar pillar;
        Geometry bottom;
        bottom = new Rectangle(12,22);
        pillar = new Pillar (bottom,58);           //pillar 是具有矩形底的柱体
        System.out.println("矩形底的柱体的体积" + pillar.getVolume());
        bottom = new Circle(10);
        pillar = new Pillar (bottom,58);           //pillar 是具有圆形底的柱体
        System.out.println("圆形底的柱体的体积" + pillar.getVolume());
    }
}
```

通过面向抽象编程设计 Pillar 类使得该 Pillar 类不再依赖具体类,因此每当系统增加新的 Geometry 的子类时,例如增加一个 Triangle 子类,不需要修改 Pillar 类的任何代码就可以使用 Pillar 创建出具有三角形底的柱体。

4.2 "开-闭"原则

所谓"开-闭"原则(Open-Closed Principle),是让设计对扩展开放,对修改关闭。怎么理解"对扩展开放,对修改关闭"呢？实际上这句话的本质是指当一个设计中增加新的模块时,不需要修改现有的模块。在给出一个设计时,应当首先考虑用户需求的变化,将应对用户变化的部分设计为对扩展开放,而设计的核心部分是经过精心考虑之后确定下来的基本结构,这部分应当是对修改关闭的,即不能因为用户的需求变化而再发生变化,因为这部分不是用来应对需求变化的。如果一个设计遵守了"开-闭"原则,那么这个设计一定是易维护的,因为在设计中

增加新的模块时,不必去修改设计中的核心模块。

4.1节的例4-1中的Geometry类和Pillar类就是系统中对修改关闭的部分,而Geometry类的子类是对扩展开放的部分。当向系统再增加任何Geometry类的子类时(对扩展开放),不必修改Pillar类,就可以使用Pillar类创建出具有Geometry类的新子类指定的底的柱体。

通常无法让设计的每个部分都遵守"开-闭"原则,甚至不应当这样去做(这样会增加系统的层次结构,降低效率),应当把主要精力集中在应对设计中最有可能因需求变化而需要改变的地方,然后想办法应用"开-闭"原则。

当设计某些系统时,我们经常需要面向抽象来考虑系统的总体设计,不考虑具体类,这样就容易设计出满足"开-闭"原则的系统。在程序设计好后,首先对abstract类的修改关闭,否则一旦修改abstract类,将可能导致它的所有子类都需要做出修改;应当对增加abstract类的子类开放,即再增加新子类时,不需要修改其他面向抽象类而设计的重要类。

在本节中,提出了"标准""面向标准的产品"和"标准组件"的新概念,这些概念更有利于理解抽象类、接口和面向抽象编程的思想。

1. 标准与接口及抽象类

"标准"是指软件产品必须具有哪些行为功能,但不指定行为功能的具体体现形式,即一个标准可以是一个抽象类或一个接口。例如,希望某些电器产品类都有on()和off()功能,那么就应该将on()和off()形成一个标准,即事先定义一个抽象类,该抽象类中定义on()和off()抽象方法,因此该抽象类并不需要具体给出on()和off()的具体实现,而on()和off()的具体行为(怎样打开和关闭电器设备)应当由该抽象类的具体子类(具体的电器产品)去实现。

在软件开发中,面对用户的需求不仅要有能力抓住用户的需求,而且更重要的是要有前瞻性,让设计的软件产品能应对用户需求的变化,即易维护、可扩展。例如,用户需要的软件产品能播放各种声音,显然在设计"声音"模拟器的类中不能只引用狗的声音,即不能用"狗"声明对象,否则编写的"声音"模拟器的类所创建的对象(软件产品)只能播放狗的声音。

如果用户对产品的某个行为功能的需求经常发生变化,就应当将这部分需求概括成一句话,即根据需求定义出一个抽象方法,并将该方法封装在接口或抽象类中,即将用户的需求形成一个标准。例如,定义播放声音的抽象方法playSound(),代码如下:

Sound.java

```java
public interface Sound {
    public abstract void playSound();
}
```

2. 面向标准的产品

现在,按照面向抽象的设计思想,只需要面向"标准",即面向接口或抽象类设计用户需要的产品。以下的Simulator类(模拟器)有一个成员变量是Sound接口类型(Simulator类面向了Sound接口),代码如下:

Simulator.java

```java
public class Simulator {
    Sound sound;
    public void setSound(Sound sound) {
        this.sound = sound;
    }
```

```java
    public void play() {
        if(sound!= null) {
            sound.playSound();
        }
        else {
            System.out.println("没有可播放的声音");
        }
    }
}
```

现在,用户程序使用Simulator类创建的对象还无法播放声音,因为还没有具体的能产生"声音"的类。例如,下列用户程序可以编译通过,但输出的结果是"没有可播放的声音"。

Application.java

```java
public class Application {
    public static void main(String args[]) {
        Simulator simulator = new Simulator();
        simulator.play();
    }
}
```

3. 标准组件

有了标准之后,就可以根据标准生产具体的"标准组件"。标准组件是标准的具体实现,即抽象类的子类或是实现接口的类。以下的Dog类(狗的声音)、Violin类(小提琴的声音)都是"标准组件"。

Dog.java

```java
public class Dog implements Sound {
    public void playSound() {
        System.out.println("狗的叫声:汪汪...汪汪");
    }
}
```

Violin.java

```java
public class Violin implements Sound {
    public void playSound() {
        System.out.println("小提琴的声音...梁祝");
    }
}
```

4. 在"开-闭"原则中的角色

"面向标准的产品"和"标准"之间的关系是组合关系,属于弱耦合关系,这有利于产品的维护和升级;而"标准组件"和"标准"之间是继承或实现关系,属于强耦合关系,即"标准组件"必须是符合标准的组件。由于"面向标准的产品"是面向抽象类或接口设计的类,因此可以使用任何一个"标准组件",即引用任何一个"标准组件"的实例。

"标准"通过接口或抽象类来体现,表明其重要性,而"标准"的具体实现由"标准组件"来负责。因此,"标准"一旦确定,就不要轻易修改,否则将导致修改所有的"标准组件"。

按照"开-闭"原则,"面向标准的产品"和"标准"都是"闭"部分,而"标准组件"是"开"部分,即"标准组件"是应对用户需求变化的部分,每当有新的"标准组件"产生,都无须修改"面向标准的产品",该产品就可以使用新的"标准组件",如图4.2所示。

5. 用户程序

以下用户程序(Application.java)播放了狗和小提琴声音。程序运行效果如图4.3所示。

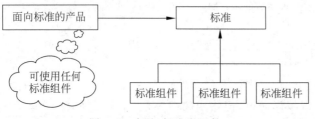

图4.2 产品、标准和组件　　　　　　　图4.3 程序运行效果

Application.java

```java
public class Application {
    public static void main(String args[]) {
        Simulator simulator = new Simulator();
        simulator.setSound(new Dog());
        simulator.play();
        simulator.setSound(new Violin());
        simulator.play();
    }
}
```

6. 构成的框架

如果将上述 Sound 接口、Simulator 类、Dog 类和 Violin 类看作一个小的开发框架,将 Application.java 看作使用该框架进行应用开发的用户程序,那么框架满足"开-闭"原则,该框架相对用户的需求比较容易维护,因为当用户程序需要模拟老虎的声音时,系统只需要简单地扩展框架,即在框架中增加一个实现 Sound 的 Tiger 类,而无须修改框架中的其他类,如图4.4所示。

图4.4 满足"开-闭"原则的框架

4.3 "多用组合,少用继承"原则

方法复用的两种最常用的技术是类的继承和对象的组合。

1. 继承与复用

子类继承父类的方法作为自己的一个方法,就好像它们是在子类中直接声明一样,可以被子类中自己声明的任何实例方法调用。也就是说,父类的方法可以被子类以继承的方式复用。

通过继承来复用父类的方法的优点是:子类可以重写父类的方法,即易于修改或扩展那些被复用的方法。

通过继承来复用方法的缺点是:

(1) 子类从父类继承的方法在编译时就确定下来了,所以无法在运行期间改变从父类继承的方法的行为。

(2) 子类和父类的关系是强耦合关系,也就是说,当父类的方法的行为更改时,必然导致子类发生变化。

(3) 通过继承进行复用也称"白盒"复用,其父类的内部细节对于子类而言是可见的。

2. 组合与复用

通过组合对象来复用方法的优点是:

(1) 通过组合对象来复用方法也称"黑盒"复用,因为当前对象只能委托所包含的对象调用其方法,这样一来,当前对象所包含的对象的方法的细节对当前对象是不可见的。

(2) 对象与所包含的对象属于弱耦合关系,因为如果修改当前对象所包含的对象的类的代码,不必修改当前对象的类的代码。

(3) 当前对象可以在运行时动态指定所包含的对象。

3. 多用组合,少用继承

之所以提倡"多用组合,少用继承",是因为在许多设计中,人们希望系统的类之间尽量是弱耦合关系,而不是强耦合关系。设计的底层结构中通常会出现较多的继承结构,而许多应用层需要避开继承的缺点,利用组合的优点。例如,在设计"中国人"时,会出现"中国人"与"人"的继承关系,当一个"中国人"的对象,例如"张三",参与活动时,例如成为一个公司的职员时,公司和"张三"应当是组合关系。

关于"多用组合,少用继承",在后面探讨重要的设计模式时,将结合中介者模式、装饰模式给予重点讲解。

4. 动态更换驾驶员

组合关系是弱耦合关系,当前对象可以在运行时动态指定所包含的对象。下面通过一个简单形象的例子——"汽车动态更换驾驶员"来体会当前对象可以在运行时动态指定所包含的对象。所谓汽车动态更换驾驶员,就是在不停车的情况下更换驾驶员,用 Java 程序来模拟,就是在不中断程序运行的情况下,给程序添加新的功能模块。

(1) 将 Driver 类和 Car 类编译通过。

Driver.java

```
public abstract class Driver {
    public abstract String getDriverLicense();
}
```

Car.java

```
public class Car {
    Driver person;                    //组合驾驶员,面向抽象
    public void setPerson(Driver p) {
```

```
            person = p;
        }
        public void show() {
            if(person == null) {
                System.out.println("目前没人驾驶汽车。");
            }
            else {
                System.out.println("目前驾驶汽车的是:");
                System.out.println(person.getDriverLicense());
            }
        }
    }
```

(2) 将主类 Application.java 编译通过，并运行起来。

Application.java

```
public class Application {
    public static void main(String arg[]) {
        Car car = new Car();
        int i = 1;
        while(true) {
            try{
                car.show();
                Thread.sleep(2000);                    //每隔2000毫秒更换驾驶员
                Class<?> cs = Class.forName("Driver" + i);
                Driver p = (Driver)cs.getDeclaredConstructor().newInstance();
                //如果没有第i个驾驶员就触发异常，跳到catch,即无人驾驶或当前驾驶员继续驾驶
                car.setPerson(p);                      //更换驾驶员
                i++;
            }
            catch(Exception exp){
                i++;
            }
            if(i>10) i = 1;                            //最多10个驾驶员轮换驾驶汽车
        }
    }
}
```

(3) 不要终止在步骤(2)运行起来的程序，继续编辑、编译 Person 类的子类。

继续编辑 Driver 类的子类，但子类的名字必须是 Driver1,Driver2,…,Driver10（顺序可任意），即单词 Driver 后跟一个不超过 10 的正整数，例如：

Driver1.java

```
public class Driver1 extends Driver {
    public String getDriverLicense() {
        return "驾驶员是赵师傅";
    }
}
```

Driver6.java

```java
public class Driver6 extends Driver {
    public String getDriverLicense() {
        return "驾驶员是钱师傅";
    }
}
```

在编辑、编译类名形如 Driver1,Driver2,…,Driver10 的 Person 类的子类时,要密切注意步骤(2)运行起来的程序的运行效果的变化(观察汽车更换的驾驶员)。这里的运行效果如图 4.5 所示。

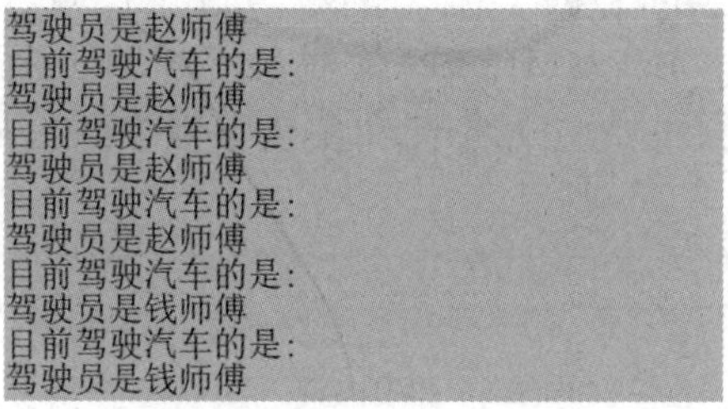

```
驾驶员是赵师傅
目前驾驶汽车的是:
驾驶员是赵师傅
目前驾驶汽车的是:
驾驶员是赵师傅
目前驾驶汽车的是:
驾驶员是赵师傅
目前驾驶汽车的是:
驾驶员是钱师傅
目前驾驶汽车的是:
驾驶员是钱师傅
```

图 4.5 程序运行效果

4.4 高内聚-低耦合原则

如果类中的方法是一组相关的行为,则称该类是高内聚的,反之称为低内聚的。高内聚有利于类的维护,而低内聚不利于类的维护。低耦合就是尽量不要让一个类含有太多的其他类的实例的引用,以避免修改系统中的一部分会影响到其他部分。

第 5 章 UML类图简介

因为本书只使用类图,所以本章简要介绍统一建模语言(Unified Modeling Language,UML)中的类图。

类图(Class Diagram)属于结构图,常被用于描述一个系统的静态结构。一个类图中通常包含类(Class)的 UML 图、接口(Interface)的 UML 图、泛化关系(Generalization)的 UML 图、关联关系(Association)的 UML 图、依赖关系(Dependency)的 UML 图和实现关系(Realization)的 UML 图。

5.1 类

在 UML 中,使用一个长方形描述一个类的主要构成,将长方形垂直地分为三层,如图 5.1 所示。

顶部第 1 层是名字层。如果类名字是常规字形,表明该类是具体类;如果类名字是斜体字形,表明该类是抽象类。

第 2 层是变量层,也称属性层,列出类的成员变量及类型,格式是"变量名字:类型"。在用 UML 表示类时,可以根据设计的需要只列出最重要的成员变量的名字。如果变量的访问权限是 public 的,需要在变量的名字前面用"+"符号修饰;如果变量的访问权限是 protected 的,需要在变量的名字前面用"♯"符号修饰;如果变量的访问权限是 private 的,需要在变量的名字前面用"-"符号修饰;如果变量的访问权限是友好的,变量的名字前面不使用任何符号修饰。

图 5.1 类的 UML 图

第 3 层是方法层,也称操作层,列出类的方法及返回类型,格式是"方法名字(参数列表):类型"。在用 UML 表示类时,可以根据设计的需要只列出最重要的方法。如果方法的访问权限是 public 的,需要在方法的名字前面用"+"符号修饰;如果方法的访问权限是 protected 的,需要在方法的名字前面用"♯"符号修饰;如果方法的访问权限是 private 的,需要在方法的名字前面用"-"符号修饰;如果方法的访问权限是友好的,方法的名字前面不使用任何符号修饰;如果方法是静态方法,方法的名字下面加下画线。

5.2 接口

UML 表示接口的 UML 图和表示类的 UML 图类似,使用一个长方形描述一个接口的主要构成,将长方形垂直地分为三层,如图 5.2 所示。

顶部第 1 层是名字层,接口的名字必须是斜体字形,而且需要用<< interface >>修饰名字,并且该修饰和名字分列在两行。

第 2 层是常量层,列出接口中的常量及类型,格式是"常量名字:类型"。在 Java 接口中,常量的访问权限都是 public 的,所以需要在常量名字前面用"+"符号修饰。

第 3 层是方法层,也称操作层,列出接口中的方法及返回类型,格式是"方法名字(参数列表):类型"。在 Java 接口中,方法的访问权限都是 public 的,所以需要在方法名字前面用"+"符号修饰。

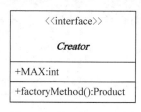

图 5.2 接口的 UML 图

5.3 泛化关系

对于面向对象语言,UML 中所说的泛化关系是指类的继承关系。如果一个类是另一个类的子类,那么 UML 通过使用一个实线连接两个类的 UML 图来表示二者之间的继承关系,实线的起始端是子类的 UML 图,终点端是父类的 UML 图,但终点端使用一个空心的三角形表示实线的结束,如图 5.3 所示。

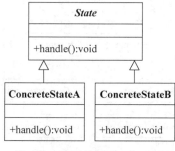

图 5.3 继承关系的 UML 图

5.4 关联关系

如果 A 类中的成员变量是用 B 类(接口)声明的变量,那么 A 和 B 的关系是关联关系(也称组合关系),称 A 关联于 B(A 组合 B)。如果 A 关联于 B,那么 UML 通过使用一个实线连接 A 和 B 的 UML 图,实线的起始端是 A 的 UML 图,终点端是 B 的 UML 图,但终点端使用一个指向 B 的 UML 图的方向箭头表示实线结束,如图 5.4 所示。

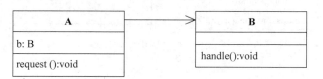

图 5.4 关联关系的 UML 图

5.5 依赖关系

如果 A 类中某个方法的参数是用 B 类(接口)声明的变量,或某个方法返回的数据类型是 B 类型的,那么 A 和 B 的关系是依赖关系,称 A 依赖于 B。如果 A 依赖于 B,那么 UML 通过使用一个虚线连接 A 和 B 的 UML 图,虚线的起始端是 A 的 UML 图,终点端是 B 的 UML

图,但终点端使用一个指向 B 的 UML 图的方向箭头表示虚线结束,如图 5.5 所示。

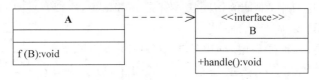

图 5.5 依赖关系的 UML 图

5.6 实现关系

如果一个类实现了一个接口,那么类和接口的关系是实现关系,称类实现接口。UML 通过使用虚线连接类和它所实现的接口,虚线起始端是类,终点端是它实现的接口,但终点端使用一个空心的三角形表示虚线的结束,如图 5.6 所示。

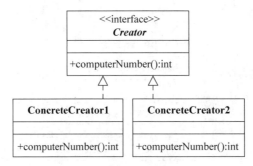

图 5.6 实现关系的 UML 图

5.7 注释

UML 使用注释为类图提供附加的说明。UML 在一个带卷角的长方形中显示给出的注释,并使用虚线将这个带卷角的长方形和它所注释的实体连接起来,如图 5.7 所示。

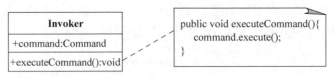

图 5.7 在类图中添加注释

第 6 章　设计模式简介

本章简要介绍设计模式,包括设计模式的起源、GoF 著作的贡献以及设计模式与框架的区别。

6.1　什么是设计模式

设计模式是针对某一类问题的最佳解决方案,而且已经被成功应用于许多系统的设计中,它解决了在某种特定情景中重复出现的某个问题。因此,可以这样定义设计模式(pattern):"设计模式是从许多优秀的软件系统中总结出的成功的、可复用的设计方案。"建筑大师 Alexander 关于设计模式的经典定义是:"每一个设计模式描述一个在我们周围不断重复出现的问题,以及该问题的解决方案的核心。这样,你就能一次一次地使用该方案而不必做重复劳动。"尽管 Alexander 所指的是城市和建筑设计模式,但他的思想也同样适用于面向对象设计模式,只是在面向对象的解决方案中,用对象和接口代替了墙壁和门窗。两类模式的核心都在于提供了相关问题的解决方案。

记录一个设计模式需有四个基本要素[1]。

1) 名称

一个模式的名称高度概括该模式的本质,有利于该行业术语统一,便于交流使用。

2) 问题

问题描述应该在何时使用模式,解释设计问题和问题存在的前因后果,描述在怎样的环境下使用该模式。

3) 方案

方案描述设计的组成部分、它们之间的相互关系及各自的职责和协作方式。

4) 效果

效果描述模式的应用效果及使用模式应当权衡的问题。效果主要包括使用模式对系统的灵活性、扩充性和复用性的影响。

例如,中介者模式的四个基本要素如下。

- 名称:中介者。
- 问题:用一个中介者封装一系列的对象交互。中介者使各对象不需要显式地相互引用,从而使其耦合松散,而且可以独立地改变它们之间的交互。
- 方案:中介者(Mediator)接口、具体中介者(Concrete Mediator)、同事(Colleague)、具体同事(Concrete Colleague)。
- 效果:减少了子类的生成,将各个同事解耦,简化了对象协议,控制集中化。

6.2　设计模式的起源

软件领域的设计模式起源于建筑学。1977 年,建筑大师 Alexander 出版了 *A Pattern Language*:*Towns*,*Building*,*Construction* 一书,Alexander 在该著作中将建筑行业中许多

问题的最佳解决方案记录为 200 多种模式，这些模式为房屋与城市的建设制定了一些规则。Alexander 著作中的思想不仅在建筑行业影响深远，而且很快影响到了软件设计领域。1987 年，受 Alexander 著作的影响，Kent Beck 和 Ward Cunningham 将 Alexander 在建筑学上的模式观点应用于软件设计，开发了一系列模式，并用 Smalltalk 语言实现了雅致的用户界面。Kent Beck 和 Ward Cunningham 在 1987 年举行的一次面向对象的会议上发表了论文《在面向对象编程中使用模式》，该论文发表后，有关软件的设计模式论文以及著作相继出版。

6.3 GoF 之书

目前，在设计模式领域公认的最具影响力的著作是 Erich Gamma、Richard Helm、Ralph Johnson 和 John Vlissides 在 1994 年合作出版的著作 *Design Patterns：Elements of Reusable Object Oriented Software*（中译本《设计模式：可复用的面向对象软件的基本原理》），该书记录了四位作者在他们四年多的工作中所发现的 23 个模式。这部著作成为空前的畅销书，对软件设计人员学习、掌握和使用设计模式产生了巨大的影响。《设计模式：可复用的面向对象软件的基本原理》一书被广大喜爱者昵称为 GoF(Gang of Four)之书，被认为是学习设计模式的必读著作，也被公认为设计模式领域的奠基之作。

自 GoF 之书出版后，受其影响，陆续出版了许多具有一定影响力的书籍，例如 1998 年，Alpert、Brown 和 Woolf 出版 *The Design Pattern Smalltalk Companion*，该书使用 Smalltalk 语言讲解了 GoF 之书中的 23 个模式；2000 年，James W. cooper 出版 *Java Design Patterns：A Tutorial*，该书使用 Java 语言讲解了 GoF 之书中的 23 个模式（中译本《Java 设计模式》），尤其侧重使用 GUI 程序设计讲解怎样使用 GoF 之书中的 23 种模式；特别要提到的是 Eric Freema 等在 2004 年出版的 *Head First Design Pattern*（中译本《Head First 设计模式》），该书使用 Java 语言重点讲解 GoF 之书中的部分模式(13 个模式)，该书图文并茂、独具匠心的写作风格令人耳目一新，语言叙述及结构安排非常适合初学者。在本书参考文献部分还列出了部分具有一定影响力的有关设计模式的著作。

6.4 学习设计模式的重要性

一个好的设计系统往往是易维护、易扩展、易复用的，有经验的设计人员或团队知道如何使用面向对象语言编写出易维护、易扩展和易复用的程序代码。《设计模式：可复用的面向对象软件的基本原理》一书正是从这些优秀的设计系统中总结出的设计精髓。尽管 GoF 之书并没有收集全部的模式(收集全部的模式似乎是不可能的，也是不必要的)，但所阐述的 23 种模式无疑是使用频率最高的模式。

使用设计模式的目的不是针对软件设计和开发中的每个问题都给出解决方案，而是针对某种特定环境中经常遇到的软件开发问题给出可重用的解决方案。因此，学习设计模式不仅可以用好这些成功的模式，而且可以更加深刻地理解面向对象的设计思想，更好地使用面向对象语言解决设计中的问题。另外，学习设计模式对于进一步学习、理解和掌握框架是非常有帮助的，例如 Java EE 中就大量使用了《设计模式：可复用的面向对象软件的基本原理》一书中的模式，对于熟悉设计模式的开发人员，很容易理解这些框架的结构，继而可以很好地使用框架设计他们的系统。《设计模式：可复用的面向对象软件的基本原理》一书所总结的成功模式不仅适合于面向对象语言,其思想及解决问题的方式也适合于任何和设计相关的行业,因此学习、掌握设计模式无疑是非常有益的。

> **注意**:Java 是面向对象语言,很多新的技术领域都涉及 Java 语言,在国内外许多大学,Java 程序设计已经成为计算机相关专业一门专业基础课。因此,采用 Java 语言讲解 GoF 之书中的设计模式非常有利于将设计模式的内容作为 Java 程序设计的后继课程。

6.5 合理使用模式

不是软件的任何部分都需要套用模式来设计,必须针对具体问题合理地使用模式。

1. 正确使用

当设计某个系统,并确认所遇到的问题刚好适合使用某个模式时,就可以考虑将该模式应用到自己的系统设计中,毕竟该模式已经被公认为解决该问题的成功方案,能使设计的系统易维护、可扩展性强、复用性好,而且这些经典的模式也容易让其他开发人员了解你的系统和设计思想。

2. 避免教条

模式不是数学公式,也不是物理定律,更不是软件设计中的"法律"条文,一个模式只是成功解决某个特定问题的设计方案,完全可以修改模式中的部分结构以符合自己的设计要求。

3. 模式挖掘

模式不是用理论推导出来的,而是从真实世界的软件系统中被发现、按照一定规范总结出来的可以被复用的方案。目前,许多文献或书籍里阐述的众多模式实际上都是 GoF 之书中经典模式的变形,这些变形模式都经过所谓的"三次规则",即该模式已经在真实世界的三个方案中被成功地采用。也可以从某个系统中寻找新模式,需要注意的是,在寻找新模式之前,必须先精通现有的模式,尤其是 GoF 之书中的 23 个模式,因为许多模式事实上只是现有模式的变种。通过研究现有的模式,不仅可以比较容易地识别模式,而且可以学会综合地使用各种模式,即使用复合模式。如果认为自己真的发现了一种新的模式,那么就可以按照 GoF 之书中提供的格式将"准模式"写成一份文档,按 6.1 节给出的模式定义,该文档需要包括名称、问题、方案和效果四个方面。当然,"准模式"需要经过"三次规则"才能成为真正的模式。

4. 避免乱用

不是所有的设计都需要使用模式,事实上,真实世界中的许多设计实例都没有使用过模式。在进行设计时,要尽可能用最简单的方法满足系统的要求,而不是费尽心机地琢磨如何在一个问题中使用模式。如果在设计中牵强地使用模式,会增加许多额外的类和对象,影响系统的性能。

5. 了解反模式

所谓反模式,是从某些软件系统中总结出的不好的设计方案。反模式就是告诉你如何采用一个不好的方案解决一个问题。既然是一个不好的方案,为何还有可能被重复使用呢?这是因为这些不好的方案表面上往往有很强的吸引力,人们很难一眼就发现它的弊端。因此,发现一个反模式也是非常有意义的工作。在有了一定的设计模式的基础之后,可以用搜索引擎查找有关反模式的信息,这对学习好设计模式也是非常有帮助的。

6.6　模式与框架

框架不是模式，框架是针对某个领域，提供用于开发应用系统的类的集合。程序设计者可以使用框架提供的类设计一个应用程序，而且在设计应用程序时可以针对特定的问题使用某个模式。

模式和框架的区别可以从以下几方面区分。

1. 层次不同

模式比框架更抽象，模式是在某种特定环境中，针对一个软件设计出现的问题给出的可复用的解决方案。模式不能向使用者提供可以直接使用的类，设计模式只有在被设计人员使用时才能表示为代码。例如，GoF 描述的中介者模式是"用一个中介对象来封装一系列的对象交互。中介者使各对象不需要显式地相互引用，从而耦合松散，而且可以独立地改变它们之间的交互"。中介者模式在解决方案中并没有提供任何类的代码，只是说明设计者可以针对特定的问题使用该模式给出的方案。框架和模式不同，它不是一种可复用的设计方案，它是由解决某个问题的一些类组成的集合，程序设计人员通过使用框架提供的类或扩展框架提供的类进行应用程序的设计。例如，在 Java 中，开发人员使用 Swing 框架提供的类设计用户界面，使用 Set(集合)框架提供的类处理数据结构相关的算法等。

2. 应用范围不同

模式本质上是逻辑概念，以概念的形式存在，模式所描述的方案独立于编程语言。Java 程序员、C++程序员或 SmallTalk 程序员都可以在自己的系统设计中使用某个模式。框架应用的范围是很具体的，它们不是以概念的形式存在，而是以具体的软件组织形式存在，只能被特定的软件设计者使用。例如，Java 提供的 Swing 框架和集合框架只能被 Java 应用程序使用。

3. 相互关系

一个框架往往包含多个设计模式，它们是面向对象系统获得最大复用的方式。较大的面向对象应用会由多层彼此合作的框架组成，例如 Java Web 设计中的 Spring 和 Hibernate 等框架。框架变得越来越普遍和重要，导致许多开源框架的出现，而且一个著名的框架往往是许多设计模式的具体体现，甚至可以在一些成功的框架中挖掘出新的模式。

第 7 章　策略模式

以下文本框中的内容引自 GoF 所著 *Design Patterns：Elements of Reusable Object Oriented Software* 的中译本及英文版。

> **策略模式**（别名：政策）
> 　　定义一系列算法，把它们一个个封装起来，并且使它们可相互替换。本模式使算法可独立于使用它的客户而变化。
> **Strategy Pattern**（Another Name：Policy）
> 　　Define a family of algorithms, encapsulate each one, and make them inter changeable. Strategy lets the algorithm vary independently from clients that use it.

以上内容是 GoF 对策略模式的高度概括，结合 7.2.1 节的策略模式的类图可以准确地理解该模式。

7.1　概述

方法是类中最重要的组成部分，一个方法的方法体由一系列语句构成，也就是说，一个方法的方法体是一个算法。在某些设计中，一个类的设计人员经常可能涉及这样的问题：由于用户需求的变化，导致经常需要修改类中某个方法的方法体，即需要不断地改变算法。

例如，有一个 Army 类，该类中有 void lineUp(int a[])方法，其中，数组 a 的元素的值代表士兵的号码，该方法将士兵按他们的号码从小到大排队，即将数组 a 按升序排列。类图如图 7.1 所示。

那么，Army 类创建的对象，例如"三连长"，调用 lineUp()方法只能将自己管理的士兵按其号码从小到大排队，如图 7.2 所示。

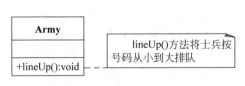

图 7.1　Army 类

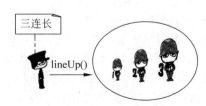

图 7.2　Army 类创建的对象调用 lineUp()方法

但有些部队希望 Army 创建的"三连长"能将士兵按照他们的号码从大到小排队，而不是从小到大排队，或将士兵按照他们的号码的某种排列来排队。显然，Army 无法提供这样的对象，如图 7.3 所示。

我们只好修改 lineUp()方法的方法体，但马上发现这样做也不行，因为一旦将 lineUp()

方法的方法体修改成把士兵按照他们的号码从大到小排队,就无法满足某些部队希望 Army 创建的"三连长"将自己的士兵从小到大排序的需求。也许可以在 lineUp()方法中添加多重条件语句,以便根据用户的具体需求决定怎样排队,但这也不是一个好办法,因为一旦有新的需求,就要修改 lineUp()方法添加新的判断语句,而且针对某个条件语句的排队代码也可能因该用户的需求变化导致重新编写。

图 7.3　Army 对象无法满足新需求

如果因为需求的变化导致经常地修改 lineUp()方法体中的代码(具体算法),这显然不利于 Army 类的维护。不用担心,面向对象编程有一个很好的设计原则——"面向抽象编程",该原则的核心是将类中经常需要变化的部分分割出来,并将每种可能的变化对应地交给抽象类的一个子类或实现接口的一个类去负责,从而让类的设计者不去关心具体实现,避免所设计的类依赖于具体的实现。基于该原则就可以使设计的类应对用户需求的变化。关于"面向抽象编程"曾在 4.1 节中讨论过,其关键有以下两点。

1. 分割变化

如果每当用户有新的需求,就会导致修改类的某部分代码,那么就应当将这部分代码从该类中分割出去,使它和类中其他稳定的代码之间是松耦合关系,即将每种可能的变化对应地交给实现某接口的类或某个抽象类的子类去负责完成。

现在,针对 Army 类中 lineUp()方法的方法体中的内容,抽象出一个"算法"标识,即一个抽象方法 abstract void arrange(int a[]),并将该抽象方法封装在一个接口或抽象类中。这里将接口命名为 LineUpStrategy,实现 LineUpStrategy 接口的类将实现接口中的 arrange(int a[])方法。例如 StrategyA 类实现 LineUpStrategy 接口,该类中的 arrange(int a[])方法把数组 a 的元素从小到大排列;StrategyB 类实现 LineUpStrategy 接口,该类中的 arrange(int a[])方法把数组 a 的元素从大到小排列;StrategyC 类实现 LineUpStrategy 接口,该类中的 arrange(int a[])方法把数组 a 的元素按奇、偶排列,并且奇数按降序,偶数按升序,例如,排列为 5,3,1,2,4,6。

2. 面向抽象设计类

现在,面向接口(抽象类)来重新设计 Army 类,让 Army 类依赖于 LineUpStrategy 接口,即 Army 类含有一个 LineUpStrategy 接口声明的变量 strategy,并重新编写 lineUp()方法的方法体中的代码,其主要代码是委托 Army 类中的 LineUpStrategy 接口变量 strategy 调用 arrange(int a[ ])方法。类图如图 7.4 所示。

如果准备让 Army 类创建的对象,例如"三连长",调用 lineUp()方法将自己的士兵从小到大排队,那么在使用 Army 类创建"三连长"时,将一个 StrategyA 类的实例的引用传递给 Army 类中的 strategy 变量;如果准备让 Army 类创建的对象,例如"三连长",调用 lineUp()方法将自己的士兵从大到小排队,那么在使用 Army 类创建"三连长"时,将一个 StrategyB 类的实例的引用传递给 Army 类中的 strategy 变量,如图 7.5 所示。

第7章 策略模式

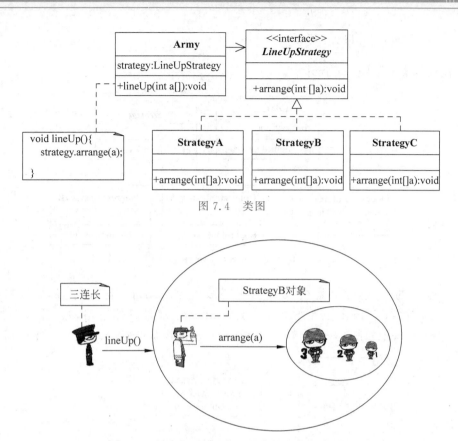

图 7.4 类图

图 7.5 实现 Strategy 接口的类的实例负责排队

当用户有新的需求时,不需要更改 Army 类的代码,只需要给出一个新的实现 LineUpStrategy 接口的类即可,该类实现的 arrange(int a[])方法能满足新需求。

策略模式是处理算法不同变体的一种成熟模式,策略模式通过接口或抽象类封装算法的标识,即在接口中定义一个抽象方法,实现该接口的类将实现接口中的抽象方法。策略模式把针对一个算法标识的一系列具体算法分别封装在不同的类中,使得各个类给出的具体算法可以相互替换。在策略模式中,封装算法标识的接口称作策略,实现该接口的类称作具体策略。

7.2 模式的结构与使用

策略模式的结构中包括三种角色。

1. 策略(Strategy)

策略是一个接口,该接口定义若干个算法标识,即定义了若干个抽象方法。

2. 具体策略(ConcreteStrategy)

具体策略是实现策略接口的类。具体策略实现策略接口所定义的抽象方法,即给出算法标识的具体算法。

3. 上下文(Context)

上下文是依赖于策略接口的类,即上下文包含策略声明的变量。上下文中提供一个方法,该方法委托策略变量调用具体策略所实现的策略接口中的方法。

扫一扫

视频讲解

7.2.1 策略模式的 UML 类图

策略模式的类图如图 7.6 所示。

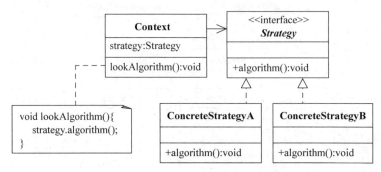

图 7.6 策略模式的类图

7.1 节提到的"三连长"是模式中的上下文角色（Army 类）的实例。策略接口是 LineUpStrategy 接口。StrategyA、StrategyB、StrategyC 都是具体策略，"三连长"可以选择 StrategyA、StrategyB、StrategyC 的某个实例作为自己的排队策略，示意图如图 7.7 所示。

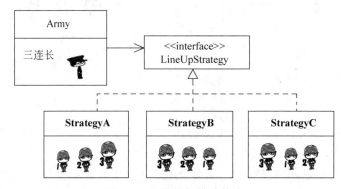

图 7.7 三连长与排队策略

7.2.2 结构的描述

下面通过一个简单的问题来描述策略模式中所涉及的各个角色。

军训时，经常需要对士兵进行列队。请给出几种列队策略，在某次训练时，可以从给出的策略中选择一种策略用于列队。

针对上述问题，使用策略模式设计若干个类。

1. 策略

本问题中，策略（Strategy）接口的名字是 LineUpStrategy，该接口规定的算法标识，即抽象方法是 abstract double arrange(int a[])。LineUpStrategy 接口的代码如下：

LineUpStrategy.java

```java
public interface LineUpStrategy{
    public abstract void arrange(int a[]);
}
```

2. 具体策略

对于本问题,有三个具体策略:StrategyA、StrategyB 和 StrategyC。

StrategyA 类中的 void arrange(int a[])方法使用选择法,把数组 a 的元素从小到大排序。StrategyB 中的 void arrange(int a[])方法使用气泡法,把数组 a 的元素从大到小排序。StrategyC 类中的 void arrange(int a[])方法把数组 a 的元素分别按奇、偶排序。

StrategyA、StrategyB 和 StrategyC 类的代码分别如下:

StrategyA.java

```java
public class StrategyA implements LineUpStrategy{
    public void arrange(int a[]) {            //选择法,从小到大排序
        for(int i = 0;i < a.length;i++){
            int index = i,j = 0;
            for(j = i + 1;j < a.length;j++){
                if(a[j]< a[index])            //从小到大排序
                    index = j;
            }
            if(index!= i) {
                int temp = a[i];
                a[i] = a[index];
                a[index] = temp;
            }
        }
    }
}
```

StrategyB.java

```java
public class StrategyB implements LineUpStrategy{
    public void arrange(int a[]) {            //起泡法,从大到小排序
        int N = a.length;
        for(int m = 0; m < N - 1;m++) {
            for(int i = 0;i < N - 1 - m;i++){
                if(a[i]< a[i + 1]){           //从大到小排序
                    int t = a[i + 1];
                    a[i + 1] = a[i];
                    a[i] = t;
                }
            }
        }
    }
}
```

StrategyC.java

```java
public class StrategyC implements LineUpStrategy{
    public void arrange(int a[]) {                    //按奇、偶分别排序
        int oddNumberAmount = 0 ;                     //奇数的个数
        for(int i = 0;i < a.length;i++){
            if(a[i] % 2!= 0)
                oddNumberAmount++;
```

```java
        int oddArray[] = new int[oddNumberAmount];              //存放奇数
        int evenArray[] = new int[a.length - oddNumberAmount];  //存放偶数
        for(int i = 0, m = 0, n = 0; i < a.length; i++){
            if(a[i] % 2 != 0) {
                oddArray[m] = a[i];
                m++;
            }
            else {
                evenArray[n] = a[i];
                n++;
            }
        }
        for(int i = 0; i < evenArray.length; i++){
            int index = i, j;
            for(j = i + 1; j < evenArray.length; j++){
                if(evenArray[j] < evenArray[index])             //从小到大排序
                    index = j;
            }
            if(index != i) {
                int temp = evenArray[i];
                evenArray[i] = evenArray[index];
                evenArray[index] = temp;
            }
        }
        int N = oddArray.length;
        for(int m = 0; m < N - 1; m++) {
            for(int i = 0; i < N - 1 - m; i++){
                if(oddArray[i] < oddArray[i + 1]){              //从大到小排序
                    int t = oddArray[i + 1];
                    oddArray[i + 1] = oddArray[i];
                    oddArray[i] = t;
                }
            }
        }
        for(int i = 0; i < oddArray.length; i++){               //复制数组
            a[i] = oddArray[i];
        }
        for(int i = 0; i < evenArray.length; i++){              //复制数组
            a[i + oddArray.length] = evenArray[i];
        }
    }
}
```

3. 上下文

上下文是 Army 类,该类包含策略声明的变量,此变量可用于保存具体策略的引用。该类中的 lineUp(int a[])方法将委托具体策略实例调用 void arrange(int a[])方法。Army 类的代码如下:

Army.java

```java
public class Army{
    LineUpStrategy strategy;                    //需要的策略接口
    public void setStrategy(LineUpStrategy strategy){
```

```
            this.strategy = strategy;
    }
    public void lineUp(int a[]){
        if(strategy!= null)
            strategy.arrange(a);
        else
            System.out.println("没有列队策略");
    }
}
```

7.2.3 模式的使用

前面已经使用策略模式给出了可以使用的类,这些类就是一个小框架,可以使用这个小框架中的类编写应用程序。

下列应用程序中,Application.java 使用了策略模式中所涉及的类,应用程序在使用策略模式时,需要创建具体策略的实例,并将其引用传递给上下文对象。程序运行效果如图 7.8 所示。

```
列队情况(从小到大):
[1, 2, 3, 4, 5, 6]
列队情况(从大到小):
[6, 5, 4, 3, 2, 1]
列队情况(奇、偶排列):
[5, 3, 1, 2, 4, 6]
```

图 7.8 程序运行效果

Application.java

```java
import java.util.Arrays;
public class Application{
    public static void main(String args[]){
        int soldierNumberOne[] = {3,1,6,2,4,5};
        int soldierNumberTwo[] = {2,5,6,3,4,1};
        int soldierNumberThree[] = {1,3,6,2,5,4};
        Army 三连长 = new Army();                              //上下文对象
        三连长.setStrategy(new StrategyA());                    //上下文对象使用策略 A
        三连长.lineUp(soldierNumberOne);
        System.out.println("列队情况(从小到大):");
        System.out.println(Arrays.toString(soldierNumberOne));
        三连长.setStrategy(new StrategyB());                    //上下文对象使用策略 B
        三连长.lineUp(soldierNumberTwo);
        System.out.println("列队情况(从大到小):");
        System.out.println(Arrays.toString(soldierNumberTwo));
        三连长.setStrategy(new StrategyC());                    //上下文对象使用策略 C
        三连长.lineUp(soldierNumberThree);
        System.out.println("列队情况(奇、偶排列):");
        System.out.println(Arrays.toString(soldierNumberThree));
    }
}
```

注意:子类可以重写(覆盖)父类的实例方法来改变该方法的行为,使子类的对象具有和父类对象不同的行为。如果考虑到系统的扩展性和复用性,就应当注意面向对象的一个基本原则:多用组合,少用继承(见 4.3 节)。策略模式采用的是组合方法,即将一个类的某个方法内容的不同变体分别封装在不同的类中,而该类仅仅依赖这些类所实现的一个共同接口。

7.3 策略模式的优点

上下文(Context)和具体策略(ConcreteStrategy)是松耦合关系。因此,上下文只知道它要使用某一个实现 Strategy 接口类的实例,但不需要知道具体是哪一个类。

策略模式满足"开-闭"原则。当增加新的具体策略时,不需要修改上下文类的代码,上下文就可以引用新的具体策略的实例。

视频讲解

7.4 应用举例——老鼠走迷宫

1. 设计要求

使用策略模式,为老鼠提供几种走迷宫策略。

2. 设计实现

1) 策略

本问题中,策略(Strategy)接口的名字是 MazeStrategy,该接口有一个抽象方法 abstract char [][] moveInMaze(char maze[][])(参数 maze 是老鼠要走的迷宫)。MazeStrategy 接口的代码如下:

MazeStrategy.java

```java
public interface MazeStrategy{
    public static char ROAD = '路';
    public static char WALL = '■';
    public static char EXIT = '出';
    public abstract char [][] moveInMaze(char maze[][]);        //走迷宫的方法
    public default char [][] copyArray(char a[][]) {
        char maze[][] = new char[a.length][a[0].length];
        for(int i = 0;i < a.length;i++) {
            for(int j = 0;j < a[0].length;j++)
                maze[i][j] = a[i][j];
        }
        return maze;
    }
}
```

2) 具体策略

对于本问题,提供了两个具体策略:StackStrategy(栈策略)和 GreedyStrategy(贪心策略)。

(1) StackStrategy 类使用栈结构的相关算法实现走迷宫。

初始化:将老鼠的出发点(入口)压入栈。

StackStrategy 的算法如下:

① 检查老鼠是否到达出口,如果到达出口,执行③,否则执行②。

② 进行弹栈操作,如果栈是空,提示无出口,执行③。如果弹栈成功,检查从栈中弹出的点是否是出口,如果是出口,提示到达出口,执行③;否则把弹出的点标记为尝试过的点(不再对尝试过的点进行压栈操作,老鼠可以直接穿越这些标记过的路点),然后把弹出的点的周围

(东西南北)的路点压入栈,但不再对尝试过的路点进行压栈操作,然后执行①。

③ 算法结束。

StackStrategy 类的代码如下:

StackStrategy.java

```java
import java.util.Stack;
import java.awt.Point;
public class StackStrategy implements MazeStrategy {
    public char [][] moveInMaze(char a[][]){
        boolean isSuccess = false;                        //是否走迷宫成功
        char maze[][] = copyArray(a);
        int rows = maze.length;
        int columns = maze[0].length;
        int x = 0;                                        //老鼠的初始位置
        int y = 0;                                        //老鼠的初始位置
        Stack<Point> stack = new Stack<Point>();
        Point point = new Point(x,y);
        stack.push(point);                                //stack 进行压栈操作
        System.out.println("老鼠到达过的位置:");
        while(isSuccess == false) {                       //未走到迷宫出口
            if(!stack.empty())
                point = stack.pop();
            else {
                System.out.printf("无法到达出口,老鼠回到入口");
                return maze;
            }
            x = (int)point.getX();
            y = (int)point.getY();
            if(maze[x][y] == EXIT) {                      //是出口
                isSuccess = true;
                maze[x][y] = '出';
                System.out.printf("\n 到达出口:(%d,%d)",x,y);
            }
            else {
                maze[x][y] = '走';                        //表示老鼠到达过该位置
                System.out.printf("(%d,%d),",x,y);
                if(y-1>=0&&(maze[x][y-1] == ROAD||maze[x][y-1] == EXIT)) {
                                                          //西是路
                    stack.push(new Point(x,y-1));         //stack 进行压栈操作
                }
                if(x-1>=0&&(maze[x-1][y] == ROAD||maze[x-1][y] == EXIT)) {
                                                          //北是路
                    stack.push(new Point(x-1,y));         //stack 进行压栈操作
                }
                if(y+1<columns&&(maze[x][y+1] == ROAD||maze[x][y+1] == EXIT)){
                                                          //东是路
                    stack.push(new Point(x,y+1));         //stack 进行压栈操作
                }
                if(x+1<rows&&(maze[x+1][y] == ROAD||maze[x+1][y] == EXIT)) {
                                                          //南是路
                    stack.push(new Point(x+1,y));         //stack 进行压栈操作
```

```
                    }
                }
            }
            return maze;
    }
}
```

(2) GreedyStrategy 类使用贪心算法实现走迷宫。

贪心算法(又称贪婪算法)是指在对问题求解时,总是做出在当前看来最好的选择。即不从整体最优上加以考虑,算法得到的是在某种意义上的局部最优解。贪心算法不是对所有问题都能得到整体最优解,关键是贪心策略的选择。

贪心算法的特点是一步一步地进行,以当前情况为基础,根据某个算法做最优选择,而不考虑各种可能的整体情况,通过每一步贪心选择,可得到局部最优解。由于贪心算法的每一步都是局部最优解,因此,如果使用贪心算法,必须判断是否得到了最优解。例如,蒙眼爬山者每次在周围选择一个最陡峭的方向前进一小步(最优选择),但是蒙眼爬山者最后爬上的山可能不是最高的山(多峰山),假设蒙眼爬山者携带了一个自动通报海拔高度的小仪器,每次都报告海拔高度,当蒙眼爬山者发现周围没有陡峭的方向可走了,报告的海拔高度恰好是想要的高度,那就找到了最优解,否则就知道自己陷入了局部最优解,无法继续下去了。贪心算法仅仅是一种思想而已,不像我们熟悉的选择法排序等算法,有明确的算法步骤。

GreedyStrategy 的算法如下:

① 检查当前路点是否是出口,即是否是最优解,如果是最优解执行③;如果不是最优解,就降低当前路点的优先度,即将当前路点的路值减 1,并检查是否陷入局部最优解,如果陷入局部最优解,执行③,否则执行②。

② 在当前路点的周围(东西南北)选出比当前路点的路值大,而且是其中最大之一的一个新路点(当前位置周围的最大的整数之一是局部最优解之一,即老鼠需要选择的下一个路点),如果找到,将新路点设置成当前路点,然后执行①;如果没有找到,当前路点不发生变化,然后执行①。

③ 算法结束。

GreedyStrategy 类的代码如下:

GreedyStrategy. java

```
public class GreedyStrategy implements MazeStrategy{
    public char [][] moveInMaze(char a[][]){
        char mazeChar[][] = copyArray(a);
        int Y = Integer.MAX_VALUE;              //最优解(迷宫出口)
        int N = Integer.MIN_VALUE;              //无解 (墙)
        int mazeInt[][] = changeToInt(mazeChar,Y,N);  // char 数组得到一个 int 二维数组
        int mousePI = 0;                        //老鼠的初始位置是左上角
        int mousePJ = 0;
        int rows;
        int columns;
        rows = mazeChar.length;
        columns = mazeChar[0].length;
        //贪心算法:当前位置周围的最大整数之一是局部最优解之一,即老鼠选择的下一个位置
        System.out.println("老鼠到达过的位置:");
```

```java
        while(mazeInt[mousePI][mousePJ]!= Y){              //如果不是最优解
            System.out.printf("( %d, %d),",mousePI,mousePJ);
            mazeChar[mousePI][mousePJ] = '走';              //表示老鼠到达过该位置
            mazeInt[mousePI][mousePJ] -- ;                 //降低优先度
            if(mazeInt[mousePI][mousePJ] == Y){
                System.out.println("陷入局部最优解");
                break;
            //因为 Integer.MIN_VALUE - 1 = Integer.MAX_VALUE,老鼠有可能陷入局部最优解
            }
            int m = mousePI;
            int n = mousePJ;
            int max = mazeInt[mousePI][mousePJ];
            //以下算法在当前位置的周围寻找最优解(最优位置)
            if(mousePJ > = 1){                              //检查西方
                if(mazeInt[mousePI][mousePJ - 1]> max){
                    max = mazeInt[mousePI][mousePJ - 1];
                    m = mousePI;
                    n = mousePJ - 1;
                }
            }
            if(mousePI > = 1){                              //北方
                if(mazeInt[mousePI - 1][mousePJ]> max){
                    max = mazeInt[mousePI - 1][mousePJ];
                    m = mousePI - 1;
                    n = mousePJ;
                }
            }
            if(mousePJ < columns - 1){                      //东方
                if(mazeInt[mousePI][mousePJ + 1]> max){
                    max = mazeInt[mousePI][mousePJ + 1];
                    m = mousePI;
                    n = mousePJ + 1;
                }
            }
            if(mousePI < rows - 1){                         //南方
                if(mazeInt[mousePI + 1][mousePJ]> max){
                    max = mazeInt[mousePI + 1][mousePJ];
                    m = mousePI + 1;
                    n = mousePJ;
                }
            }
            mousePI = m;                                    //改变老鼠的位置
            mousePJ = n;
        }
        mazeChar[mousePI][mousePJ] = '出';
        System.out.printf("到达出口:( %d, %d)\n",mousePI,mousePJ);
        System.out.println("\n找到最优解: " + mazeInt[mousePI][mousePJ]);
        return mazeChar;
    }
    private int [][] changeToInt(char[][] mazeChar,int Y,int N){
        int rows = mazeChar.length;
```

```
            int columns = mazeChar[0].length;
            int mazeInt[][] = new int[rows][columns];
            for(int i = 0;i < rows;i++){
                for(int j = 0;j < columns;j++) {
                    if(mazeChar[i][j] == ROAD) {
                        mazeInt[i][j] = Y - 1;
                    }
                    else if(mazeChar[i][j] == WALL){
                        mazeInt[i][j] = N;
                    }
                    else if(mazeChar[i][j] == EXIT){
                        mazeInt[i][j] = Y;
                    }
                }
            }
            return mazeInt;
        }
    }
```

注意：如果路值不会被减到等于墙的值(Integer.MIN_VALUE)，就一定能到达出口，否则某个墙或路会被当成出口(因为 Integer.MIN_VALUE－1＝Integer.MAX_VALUE，老鼠陷入局部最优解。

3) 上下文

上下文是 Mouse 类，该类包含 MazeStrategy 声明的变量(组合了策略)。Mouse 类的代码如下：

Mouse.java

```
public class Mouse{
    MazeStrategy strategy;
    public void setStrategy(MazeStrategy strategy){
        this.strategy = strategy;
    }
    public char[][] move(char a[][]){
        char maze[][] = null;
        if(strategy!= null)
            maze = strategy.moveInMaze(a);
        else {
            System.out.println("没有走迷宫的策略");
        }
        return maze;
    }
}
```

4) 应用程序

下列应用程序(Application.java)使用了策略模式中所涉及的类，实现老鼠使用两种策略走迷宫。程序运行效果如图 7.9 所示。

```
老鼠要走的迷宫:
[路, 路, ■, ■, 路]
[路, 路, 路, 路, 路]
[■, 路, ■, ■, 路]
[路, 路, ■, 出, 路, ■]
使用栈策略:
老鼠到达过的位置:
(0,0), (1,0), (1,1), (2,1), (3,1), (3,0), (1,2), (1,3), (1,4), (2,4), (3,4)
到达出口:(3,3)
"走"和"出"表示老鼠曾走过的路和到达的出口:
[走, 路, ■, ■, 路]
[走, 走, 走, 走, 走]
[■, 走, ■, ■, 走]
[路, 路, ■, 出, 走, ■]
使用贪心策略:
老鼠到达过的位置:
(0,0), (1,0), (1,1), (2,1), (3,1), (3,0), (3,0), (3,1), (2,1), (1,1), (0,1), (0,1), (0,0), (1,0),
(1,0), (0,0), (0,1), (1,1), (1,2), (1,3), (1,4), (2,4), (3,4), 到达出口:(3,3)

找到最优解: 2147483647
"走"和"出"表示老鼠曾走过的路和到达的出口:
[走, 走, ■, ■, 路]
[走, 走, 走, 走, 走]
[■, 走, ■, ■, 走]
[走, 走, ■, 出, 走, ■]
```

图 7.9 程序运行效果

Application.java

```java
import java.util.Arrays;
public class Application{
    public static void main(String args[]){
        final char 路 = MazeStrategy.ROAD;    //为了让迷宫数组的编辑更加直观
        final char 墙 = MazeStrategy.WALL;
        final char 出 = MazeStrategy.EXIT;
        char maze[][] = {{ 路,路,墙,墙,路 },
                         { 路,路,路,路,路 },
                         { 墙,路,墙,墙,路 },
                         { 路,路,墙,出,路,墙 }};
        System.out.println("老鼠要走的迷宫: ");
        for(int i = 0;i<maze.length;i++){
            System.out.println(Arrays.toString(maze[i]));
        }
        Mouse jerry = new Mouse();
        System.out.println("使用栈策略:");
        jerry.setStrategy(new StackStrategy());
        char a[][] = jerry.move(maze);
        System.out.println("\n\"走\"和\"出\"表示老鼠曾走过的路和到达的出口:");
        for(int i = 0;i<a.length;i++){
            System.out.println(Arrays.toString(a[i]));
        }
        System.out.println("使用贪心策略:");
        jerry.setStrategy(new GreedyStrategy());
        a = jerry.move(maze);
        System.out.println("\"走\"和\"出\"表示老鼠曾走过的路和到达的出口:");
        for(int i = 0;i<a.length;i++){
            System.out.println(Arrays.toString(a[i]));
        }
    }
}
```

第 8 章　责任链模式

以下文本框中的内容引自 GoF 所著 *Design Patterns：Elements of Reusable Object Oriented Software* 的中译本及英文版。

> **责任链模式**
> 　　使多个对象都有机会处理请求，从而避免请求的发送者和接收者之间发生耦合关系。将这些对象连成一条链，并沿着这条链传递该请求，直到有一个对象处理它为止。
> **Chain of Responsibility Pattern**
> 　　Avoid coupling the sender of a request to its receiver by giving more than one object a chance to handle the request. Chain the receiving objects and pass the request along the chain until an object handles it.

以上内容是 GoF 对责任链模式的高度概括，结合 8.2.1 节的责任链模式的类图可以准确地理解该模式。

8.1　概述

　　在设计 Java 程序时，可能需要多个对象按照顺序响应用户的请求。例如，要建立一个古瓷器鉴定系统，一个好的设计方案是将古瓷器分门别类，然后创建若干对象，每个对象负责处理一类古瓷器的鉴定。为了能更好地组织这些负责鉴定古瓷器的对象，可以将这些对象组成一个责任链。当用户需要鉴定古瓷器时，系统可以让责任链上的第一个对象来处理用户的请求（也可以不是第一个，这依赖于具体应用）。这个对象首先检查自己是否能处理用户的请求，如果能处理就反馈有关处理结果，如果不能处理就将用户的请求传递给责任链上的下一个对象，以此类推，直到责任链上的某个对象能处理用户的请求。如果责任链上的末端对象也不能处理用户的请求，那么用户的本次请求就无任何结果。具体地，可以创建专门负责鉴定"唐代瓷""宋代瓷""明代瓷"和"清代瓷"的对象，分别命名为 A、B、C 和 D，这四个对象形成一个责任链 A→B→C→D。当用户请求鉴定自己的古瓷器时，系统将用户的请求提交给责任链上的 A 对象，如果责任链上的 A 对象无法给出处理结果，就把用户的请求传递给责任链上的 B 对象；如果 B 对象也无法给出处理结果，就把用户的请求传递给责任链上的 C 对象；如果 C 对象给出了鉴定结果"用户的古瓷器是明代瓷"，就不再把请求传递给责任链上的 D 对象。鉴定古瓷器的责任链如图 8.1 所示。

　　责任链模式的关键是将用户的请求分派给许多对象，这些对象被组织成一个责任链，即每个对象含有后继对象的引用，并要求责任链上的每个对象如果能处理用户的请求，就做出处理，不再将用户的请求传递给责任链上的下一个对象，如果不能处理用户的请求，就必须将用户的请求传递给责任链上的下一个对象。

第 8 章 责任链模式

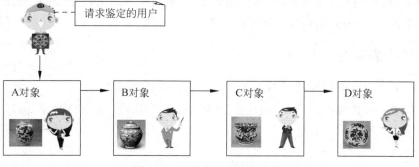

图 8.1 鉴定古瓷器的责任链

8.2 模式的结构与使用

责任链模式的结构中包括两种角色。

1. 处理者（Handler）

处理者是一个接口，负责规定具体处理者需要实现的方法，以及具体处理者设置后继者的方法。

2. 具体处理者（ConcreteHandler）

具体处理者是实现处理者接口的类，其实例是责任链上的对象。责任链上的对象通过调用处理者接口规定的方法处理用户的请求，即在接到用户的请求后，责任链上的对象将调用接口规定的方法，在执行该方法的过程中，如果发现能处理用户的请求，就处理有关数据，否则就将无法处理的信息反馈给用户，然后将用户的请求传递给自己的后继对象。

> **注意**：由于 Java 不支持多重继承，因此在 Java 中，处理者最好不是一个抽象类，否则创建具体抽象者的类将无法继承其他类，限制了具体处理者的能力。

▶ 8.2.1 责任链模式的 UML 类图

责任链模式的类图如图 8.2 所示。

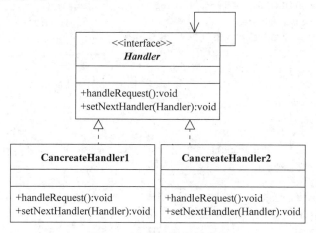

图 8.2 责任链模式的类图

8.1节提到的专门负责鉴定"唐代瓷""宋代瓷""明代瓷"和"清代瓷"的对象相当于模式中某个具体处理者(ConcreteHandler)角色的实例。

8.2.2 结构的描述

下面通过一个简单的问题描述责任链模式中所涉及的各个角色。

用户想知道一个古瓷器是不是"唐代瓷""宋代瓷""明代瓷"或"清代瓷"。针对该问题,用责任链模式模拟为用户鉴定古瓷器。

1) 处理者

在本问题中,处理者(Handle)接口的名字是Handler,代码如下:

Handler.java

```java
public interface Handler{
    public abstract void handleRequest(String number);
    public abstract void setNextHandler(Handler handler);
}
```

2) 具体处理者

本问题有两个具体处理者,即鉴定者的类,分别是ChinawareHandlerA和ChinawareHandlerB。ChinawareHandlerA类的实例,即鉴定者,使用Java的对话框(JDialog的实例)观看瓷器的照片(模拟鉴定者观察瓷器),并给出鉴定结论。如果无法给出鉴定结论,就让下一位鉴定者鉴定该瓷器。如果发现没有下一位鉴定者,就提示责任链没有给出鉴定结论。ChinawareHandlerB类的实例,即鉴定者,使用本地系统提供的"画笔"软件观看瓷器的照片(模拟鉴定者观察瓷器),并给出鉴定结论。如果无法给出鉴定结论,就让下一位鉴定者鉴定该瓷器。如果发现没有下一位鉴定者,就告知责任链没有给出鉴定结论。

ChinawareHandlerA类和ChinawareHandlerB类的代码分别如下:

ChinawareHandlerA.java

```java
import javax.swing.*;
import java.awt.*;
import java.util.*;
public class ChinawareHandlerA extends JDialog implements Handler{
    private Handler nextHandler;                         //存放当前处理者后继的Hander接口变量
    public final static String NORESULT = "未给出鉴定结论";
    String name;                                         //鉴定者的名字
    String backMess = NORESULT ;
    Image image;                                         //瓷器的图像
    ChinawareHandlerA(){
        name = "匿名";
    }
    ChinawareHandlerA(String name){
        this.name = name;
    }
    public void setNextHandler(Handler handler){         //实现处理者接口中的抽象方法
        nextHandler = handler;
    }
    public String handleRequest(String imageName){       //实现处理者接口中的抽象方法
```

第8章 责任链模式

```java
        Toolkit toolkit = Toolkit.getDefaultToolkit();
        image = toolkit.getImage(imageName);
        repaint();
        setDefaultCloseOperation(JFrame.DISPOSE_ON_CLOSE);
        setTitle(name);
        setBounds(12,12,500,600);
        setVisible(true);
        System.out.print("请"+ name +"输入鉴定结论(如果无法鉴定,请输入 0):");
        Scanner scanner = new Scanner(System.in);
        backMess = scanner.nextLine();
        if(backMess.equals("0")){
            dispose();
            if(nextHandler!= null){
                backMess = name + NORESULT;                  //没有给出鉴定结论
                String s = nextHandler.handleRequest(imageName);   //下一位处理者继续鉴定
                backMess = backMess + "." + s;
            }
            else {
                backMess = "责任链" + NORESULT;
            }
        }
        else {
            dispose();
            backMess = name + "给出的鉴定结论:" + backMess;
        }
        return backMess;
    }
    public void paint(Graphics g ) {
        super.paint(g);
        g.drawImage(image,0,0,getBounds().width,getBounds().height,this);
    }
}
```

ChinawareHandlerB.java

```java
import java.io.*;
import java.util.*;
public class ChinawareHandlerB implements Handler{
    private Handler nextHandler;                    //存放当前处理者后继的Hander接口变量
    public final static String NORESULT = "未给出鉴定结论";
    String name;                                    //鉴定者的名字
    String backMess = NORESULT ;
    ChinawareHandlerB(){
        name = "匿名";
    }
    ChinawareHandlerB(String name){
        this.name = name;
    }
    public void setNextHandler(Handler handler){    //实现处理者接口中的抽象方法
        nextHandler = handler;
    }
```

```java
public String handleRequest(String imageName){          //实现处理者接口中的抽象方法
    Runtime ce = null;
    File file = null;
    try{                                                //用本地"画笔"程序查看图像
       ce = Runtime.getRuntime();
       file = new File("mspaint.exe " + imageName);
       ce.exec(file.getName());
    }
    catch(Exception e) {}
    System.out.print("请" + name + "输入鉴定结论(如果无法鉴定,请输入 0):");
    Scanner scanner = new Scanner(System.in);
    backMess = scanner.nextLine();
    if(backMess.equals("0")){
       try{
          ce.exec("taskkill /IM mspaint.exe");          //关闭 mspaint.exe(画笔)
       }
       catch(Exception e) {}
       if(nextHandler!= null){
          backMess = name + NORESULT;                   //没有给出鉴定结论
          String s = nextHandler.handleRequest(imageName);    //下一位处理者继续鉴定
          backMess = backMess + "。" + s;
       }
       else {
          backMess = name + NORESULT + "。责任链" + NORESULT;
       }
    }
    else {
       try{
          ce.exec("taskkill /IM mspaint.exe");          //关闭 mspaint.exe(画笔)
       }
       catch(Exception e) {}
       backMess = name + "给出的鉴定结论:" + backMess;
    }
    return backMess;
 }
}
```

▶ 8.2.3 模式的使用

前面已经使用责任链模式给出了可以使用的类,这些类就是一个小框架,可以使用这个小框架中的类编写应用程序。

下列应用程序(Application.java)使用了责任链模式,创建责任链,并指定从责任链上的哪个对象开始响应用户,对用户提交的瓷器进行鉴定,给出鉴定结论。程序运行效果如图 8.3 所示。

Application.java

```java
public class Application{
    public static void main(String args[]){
        String imageName = "china.jpg";
```

第 8 章 责任链模式

(a) 责任链在鉴定瓷器

请唐代瓷器鉴定专家输入鉴定结论(如果无法鉴定,请输入0):0
请宋代瓷器鉴定专家输入鉴定结论(如果无法鉴定,请输入0):0
请明代瓷器鉴定专家输入鉴定结论(如果无法鉴定,请输入0):属于明代瓷器
用户得到的结果:
唐代瓷器鉴定专家未给出鉴定结论。宋代瓷器鉴定专家未给出鉴定结论。明代瓷器鉴定专家给出的鉴定结论:属于明代瓷器

(b) 责任链鉴定瓷器结束

图 8.3 程序运行效果

```java
        Chain chinaChain = new Chain();
        String result = chinaChain.responseClient(imageName);
        System.out.println("用户得到的结果:");
        System.out.println(result);
    }
}
class Chain {              //责任链
    ChinawareHandlerA one;
    ChinawareHandlerB two;
    ChinawareHandlerA three;
    ChinawareHandlerB four;
    public Chain(){
        four = new ChinawareHandlerB("清代瓷器鉴定专家");
        three = new ChinawareHandlerA("明代瓷器鉴定专家");
        two = new ChinawareHandlerB("宋代瓷器鉴定专家");
        one = new ChinawareHandlerA("唐代瓷器鉴定专家");
        one.setNextHandler(two);
        two.setNextHandler(three);
        three.setNextHandler(four);
        four.setNextHandler(null);
    }
    public String responseClient(String imageName){
        String s = one.handleRequest(imageName);
        return s;
    }
}
```

8.3 责任链模式的优点

责任链模式具有以下优点:

(1) 责任链中的对象只和自己的后继是低耦合关系,和其他对象毫无关联,这使得编写处理者对象以及创建责任链变得非常容易。

(2) 当在处理者中分配职责时,责任链给应用程序更多的灵活性。

(3) 应用程序可以动态地增加、删除处理者或重新指派处理者的职责。
(4) 应用程序可以动态地改变处理者之间的先后顺序。
(5) 使用责任链的用户不必知道处理者的信息,用户不会知道到底是哪个对象处理了它的请求。

8.4 应用举例——现金找零

提到找零钱,也许读者在学习某门编程语言时,用这个问题做过编程的训练。例如,用 10 元、5 元、2 元和 1 元面值的货币实现找零 28 元,共有多少种方案?在实际生活中,例如超市的现金收银员,并不是首先考虑有多少种找零的方案,然后从中选择其一。假设收银员可以使用 10 元、5 元、2 元和 1 元面值的货币完成找零,收银员找零 28 元的实际过程如下:

首先在 10 元面值的钱盒中看能否完成全部或部分任务。发现 10 元面值的钱盒能完成找零 28 元中的 20 元(贡献 2 张 10 元面值的货币),即只完成部分任务,那么就要把剩余的找零 8 元的任务交给下一个 5 元面值的钱盒。5 元面值的钱盒发现能完成找零 8 元中的 5 元(贡献 1 张 5 元面值的货币),然后把剩余的找零 3 元任务交给下一个 2 元面值的钱盒。以此类推,如果后续某个钱盒完成了找零任务,那么收银员就完成了找零任务。如果一直到最后一个钱盒(1 元面值的钱盒),该钱盒也无法完成找零任务,那么收银员最终就无法完成找零任务。为了简化算法,假设收银员始终有充足的 10 元、5 元、2 元和 1 元面值的货币,即收银员一定能完成找零任务。

1. 设计要求

使用责任链模式模拟超市收银员的找零过程。

2. 设计实现

使用责任链模拟收银员找零,那么责任链上的对象就是钱盒,即责任链模式中的具体处理者的实例。

1) 处理者

在本问题中,处理者(Handle)接口的名字是 MoneyHandler,代码如下:

MoneyHandler.java

```
public interface MoneyHandler {
    //handleChange()方法把整钱 money 分解成小于或等于 money 的零钱,并返回该零钱
    public abstract int handleChange(int money);
    public abstract void setNextMoneyHandler(MoneyHandler handler);
}
```

2) 具体处理者

对于本问题,有一个负责创建具体处理者,即负责找零的 MoneyBox 类。MoneyBox 类的代码如下:

MoneyBox.java

```
public class MoneyBox implements MoneyHandler{
    int moneyValue;                          //钱的面值
    private MoneyHandler nextHandler;        //存放当前处理者的后继处理者
    public MoneyBox(int moneyValue){
        this.moneyValue = moneyValue;
```

```java
    }
    public MoneyBox(){
        moneyValue = 1;
    }
    //用面值 moneyValue 的钱把整钱 money 分解成小于或等于 money 的零钱
    public int handleCharge(int money){
        int backMoney = 0;                        //方法的返回值,即完成的找零
        int completedChargeTasks = 0;             //本钱盒贡献的零钱
        int remainingChargeTasks = 0;             //未能完成的找零
        int n = 0,sum = 0;
        completedChargeTasks = money/moneyValue * moneyValue;
        remainingChargeTasks = money % moneyValue;
        if(completedChargeTasks == money) {       // 找零成功
            backMoney = completedChargeTasks ;
            System.out.println(
              "面值是" + moneyValue + "元的钱盒完成找零:" + completedChargeTasks + "元");
        }
        else {
            if(nextHandler != null){
                System.out.println(
                  "面值是" + moneyValue + "元的钱盒完成找零:" + completedChargeTasks + "元");
                //让下一个钱盒(处理者)完成剩余的找零任务:remainingChargeTasks
                backMoney =
                  completedChargeTasks + nextHandler.handleCharge(remainingChargeTasks);
            }
            else {
                System.out.println("找零失败");
                backMoney = 0;                    // 找零失败
            }
        }
        return backMoney;
    }
    public void setNextMoneyHandler(MoneyHandler nextHandler){
        this.nextHandler = nextHandler;
    }
}
```

3)应用程序

下列应用程序(Application.java)使用了责任链模式,创建责任链,并指定从责任链上的哪个对象开始响应用户,实现现金找零。程序运行效果如图 8.4 所示。

```
需要找零59元
面值是10元的钱盒完成找零:50元
面值是5元的钱盒完成找零:5元
面值是2元的钱盒完成找零:4元
用户得到的零钱:59
```

图 8.4 程序运行效果

Application.java

```java
public class Application{
    public static void main(String args[]){
        int money = 59;
        Chain moneyBoxChain = new Chain();
        System.out.println("需要找零" + money + "元");
        int result = moneyBoxChain.giveCharge(money);
        System.out.print("用户得到的零钱:" + result);
```

```java
        }
    }
    class Chain {                              //责任链
        MoneyBox money10,                      //10元面值的钱盒
                money5,                        //5元面值的钱盒
                money2,                        //2元面值的钱盒
                money1;                        //1元面值的钱盒
        public Chain(){
            money1 = new MoneyBox(1);
            money2 = new MoneyBox(2);
            money5 = new MoneyBox(5);
            money10 = new MoneyBox(10);
            money10.setNextMoneyHandler(money5);
            money5.setNextMoneyHandler(money2);
            money2.setNextMoneyHandler(money1);
            money1.setNextMoneyHandler(null);
        }
        public int giveCharge(int money){
            int charge = money10.handleCharge(money);
            return charge;                     //返回给用户的零钱
        }
    }
```

第 9 章 访问者模式

以下文本框中的内容引自 GoF 所著 *Design Patterns*：*Elements of Reusable Object Oriented Software* 的中译本及英文版。

> **访问者模式**
> 表示一个作用于某对象结构中的各个元素的操作。它可以在不改变各个元素的类的前提下定义作用于这些元素的新操作。
>
> **Visitor Pattern**
> Represent an operation to be performed on the elements of an object structure. Visitor lets you define a new operation without changing the classes of the elements on which it operates.

以上内容是 GoF 对访问者模式的高度概括，结合 9.2.1 节的访问者模式的类图可以准确地理解该模式。

9.1 概述

编写类的时候，在该类中也编写了若干个实例方法，该类的对象通过调用这些实例方法操作其成员变量表明所产生的行为。在某些设计中，需要定义作用于成员变量的新操作，而且这个新的操作不应当由该类中的某个实例方法来承担。例如，电表（记录用户用电情况的仪器）可以显示用电量，但不提供计费功能，根据电表的用电量进行计费的任务由电力部门负责，即将数据的存储和数据的处理解耦，这有利于对象的维护和复用，因为如果用户购买的电表本身有计费功能，一旦计费算法发生变化，必然需要修改电表，这降低了可维护性；这样的电表只适合某些用户，这降低了电表的复用性。

但电费是一定要计算的，即根据电表显示的用电量收取用户的电费。在实际生活中，由电力部门的"计表员"查看电表的用电量，然后按照有关收费标准计算出电费。

访问者模式建议让一个称作访问者的对象访问元素（电表），然后用某种算法操作元素的数据（电表的电量）。

把一栋楼比作一个集合，那么电表就是集合中的元素，计表员查看（访问）楼房中的电表，并根据电表显示的数字计算电费，如图 9.1 所示。

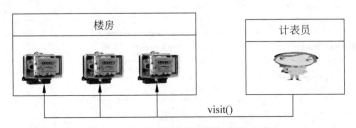

图 9.1 计表员"访问"楼房中的电表

9.2 模式的结构与使用

访问者模式包括五种角色。

1. 抽象元素（Element）

抽象元素是一个抽象类，该类定义了接收访问者的抽象方法，例如 accept（Visitor visitor）。

2. 具体元素（Concrete Element）

具体元素是抽象元素的子类。具体元素的实例（具体元素创建的对象）需要访问者帮助处理它的数据，即将数据存储与数据处理解耦，因此，具体元素必须实现抽象元素规定的接收访问者的方法，例如 accept(Visitor visitor)方法，以便指定具体的访问者。

3. 对象结构（Object Structure）

对象结构是一个集合，用于存放 Element 对象，提供遍历它自己的方法。该角色不是必需的，当需求中只有一个具体元素的实例时，就不需要该角色。

4. 抽象访问者（Visitor）

抽象访问者是一个接口，该接口定义操作对象（具体元素的实例）的方法，例如 visitor(Element element)。

5. 具体访问者（Concrete Visitor）

具体访问者实现 Visitor 接口的类。具体访问者的实例（具体访问者创建的对象）需要访问具体元素的实例，以便处理具体元素的实例的数据，因此，具体访问者必须实现抽象访问者规定的操作对象的方法，例如 visitor(Element element)方法，以便访问具体元素的实例，并处理该实例的数据。

9.2.1 访问者模式的 UML 类图

访问者模式的类图如图 9.2 所示。

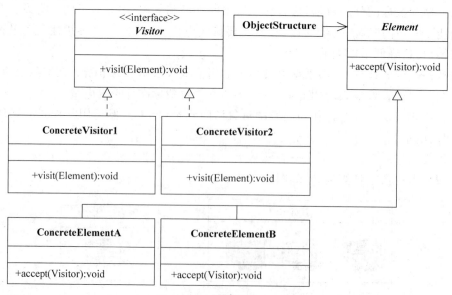

图 9.2 访问者模式的类图

9.1节中提到的"计表员"是访问者模式中的具体访问者的实例,"楼房"是对象结构的实例,"电表"是具体元素的实例。

▶ 9.2.2 结构的描述

下面通过一个简单的问题来描述访问者模式中所涉及的各个角色。

某一栋楼安装有若干个电表,开始时,电力部门是按民用电收费,后来楼内住户性质发生变化,电力部门开始按工业用电收费。

针对上述问题,使用访问者模式设计若干个类,模拟根据电表收取电费。

1) 抽象元素

抽象元素角色是 AmmeterElement 抽象类,在 AmmeterElement 类中,electricAmount 成员变量的值表示用电量,showElectricAmount()方法返回 electricAmount 变量的值。按照对本角色的要求,AmmeterElement 抽象类定义了接收访问者的抽象方法 accept(Visitor visitor)。AmmeterElement 抽象类的代码如下:

AmmeterElement.java

```java
public abstract class AmmeterElement {
    int electricAmount ;                        //用电量
    public abstract void accept(Visitor v);     //接收访问者
    public int showElectricAmount() {
       return electricAmount;
    }
    public void setElectricAmount(int n){
       electricAmount = n;
    }
}
```

2) 具体元素

对于本问题,有一个具体元素,即 Ammeter 类。Ammeter 类的实例用来模拟电表。Ammeter 类的代码如下:

Ammeter.java

```java
public class Ammeter extends AmmeterElement{
    public void accept(Visitor visitor){
        int electricityFees = visitor.visit(this);          //让访问者访问当前元素
        System.out.println("交纳电费:" + electricityFees + "元");
    }
}
```

3) 对象结构

在本问题中,该角色是 java.util 包中的 HashSet 集合。

4) 抽象访问者

抽象访问者是 Visitor 接口,按照对本角色的要求,该接口定义了 visitor(AmmeterElement element)方法。Visitor 接口的代码如下:

Visitor.java

```java
public interface Visitor{
    public int visit(AmmeterElement element);
}
```

5）具体访问者

本角色有两个类，分别是 HomeAmmeterVisitor 类和 IndustryAmmeterVisitor 类，二者都实现了 Visitor 接口。HomeAmmeterVisitor 类的实例模拟根据家用电标准收费的计表员；IndustryAmmeterVisitor 类的实例模拟根据工业用电标准收费的计表员。HomeAmmeterVisitor 类和 IndustryAmmeterVisitor 类都重写了接口的 visitor(AmmeterElement element) 方法，以便访问具体元素 Ammeter 类的实例（电表），并根据所访问的实例的数据 electricAmount（用电量）计算出电费，但二者的算法不同，即体现不同的收费标准。

HomeAmmeterVisitor 类和 IndustryAmmeterVisitor 类的代码如下：

HomeAmmeterVisitor.java

```java
public class HomeAmmeterVisitor implements Visitor{
    public int visit(AmmeterElement ammeter){
        int charge = 0;
        int unitOne = 3, unitTwo = 6;
        int basic = 6000;
        int n = ammeter.showElectricAmount();
        if(n <= basic) {
            charge = n * unitOne;
        }
        else {
            charge = basic * unitOne + (n - basic) * unitTwo;
        }
        return charge;
    }
}
```

IndustryAmmeterVisitor.java

```java
public class IndustryAmmeterVisitor implements Visitor{
    public int visit(AmmeterElement ammeter){
        int charge = 0;
        int unitOne = 5, unitTwo = 12;
        int basic = 15000;
        int n = ammeter.showElectricAmount();
        if(n <= basic) {
            charge = n * unitOne;
        }
        else {
            charge = basic * unitOne + (n - basic) * unitTwo;
        }
        return charge;
    }
}
```

▶ 9.2.3 模式的使用

前面已经使用访问者模式给出了可以使用的类，这些类就是一个小框架，可以使用这个小框架中的类编写应用程序。

在应用程序中，程序让具体元素的实例调用抽象元素规定的接收访问者的方法，例如

accept(Visitor visitor),即如果一个对象需要其他对象帮助处理自己的数据,那么这个对象需要调用方法,让能帮助自己处理数据的对象访问自己。

下列应用程序(Application.java)使用了访问者模式,模拟根据电表收取电费。程序运行效果如图9.3所示。

按民用电,各个电表的电费如下:
缴纳电费:57元
缴纳电费:36元
缴纳电费:168元
按工业用电,各个电表的电费如下:
缴纳电费:95元
缴纳电费:60元
缴纳电费:280元

图9.3 程序运行效果

Application.java

```java
import java.util.*;
public class Application{
    public static void main(String args[]) {
        Visitor chargingHome = new HomeAmmeterVisitor();           //家用电标准计算电费的"计表员"
        Visitor chargingIndustry = new IndustryAmmeterVisitor();   //工业用电标准计算电费的"计表员"
        HashSet<Ammeter> set = new HashSet<Ammeter>();             //模拟存放电表的集合
        Ammeter ammeter1 = new Ammeter(),
                ammeter2 = new Ammeter(),
                ammeter3 = new Ammeter();
        ammeter1.setElectricAmount(12);
        ammeter2.setElectricAmount(56);
        ammeter3.setElectricAmount(19);
        set.add(ammeter1);
        set.add(ammeter2);
        set.add(ammeter3);                                         //set中有三个电表
        System.out.println("按民用电,各个电表的电费如下:");
        Iterator<Ammeter> iter = set.iterator();
        while(iter.hasNext()){
            Ammeter ammeter = iter.next();
            //用accept()方法让chargingHome访问ammeter对象
            ammeter.accept(chargingHome);
        }
        System.out.println("按工业用电,各个电表的电费如下:");
        iter = set.iterator();
        while(iter.hasNext()){
            Ammeter ammeter = iter.next();
            //用accept()方法让chargingIndustry访问ammeter对象
            ammeter.accept(chargingIndustry);
        }
    }
}
```

注意:访问者模式在不改变类的情况下可以有效地增加其上的操作,为了达到这样的效果,使用了一种称为"双重分派"的技术:在访问者模式中,被访问者,即Element角色element,首先调用accept(Visitor visitor)方法"接收"访问者,而被接收的访问者visitor再调用visit(Element element)方法访问当前element对象。

9.3 访问者模式的优点

可以在不改变一个集合中元素的类的情况下,增加施加于该元素的新操作。可以将集合中各个元素的某些操作集中到访问者中,这不仅便于集合的维护,也有利于集合中元素的复用。

9.4 应用举例——答卷与批卷

1. 设计要求

试卷负责存储和显示试题；试卷需要由学生给出解答（试卷不能解答试题）；学生解答完毕后，由老师批卷并给出分数（试卷不能给出分数）。使用访问者模式给出若干个类，模拟学生答卷、老师批卷。

2. 设计实现

1) 抽象元素

在本问题中，抽象元素角色是 TestPaper 抽象类。代码如下：

TestPaper.java

```java
public abstract class TestPaper{
    public String name;
    public void setName(String name){
        this.name = name;
    }
    public abstract String [] backQuestions();        //返回试题
    public abstract int [] backAnswer();              //返回解答
    public abstract void accept(Visitor v);           //设置阅卷者或答卷者
}
```

2) 具体元素

对于本问题，提供了一个具体元素类 ArithmeticPaper，该类的实例是一张算术试卷。代码如下：

ArithmeticPaper.java

```java
public class ArithmeticPaper extends TestPaper {         //算术试卷
    String question [];                                   //存放试题
    int answerCar [];                                     //存放解答(答题卡)
    ArithmeticPaper(){
        question = new String[4];
        answerCar = new int[4];
        question[0] = "6+12 等于几?";
        question[1] = "6+2×5 等于几?";
        question[2] = "30-10÷5 等于几?";
        question[3] = "28％5 等于几?";
    }
    public String [] backQuestions(){
        return question;
    }
    public int [] backAnswer(){
        return answerCar;
    }
    public void accept(Visitor v){
        v.visit(this);
    }
}
```

3）对象结构

对象结构角色是一个数组。

4）抽象访问者

在本问题中，抽象访问者是 Visitor 接口。代码如下：

Visitor. java

```java
public interface Visitor {
    public void visit(TestPaper paper);  //访问试卷
}
```

5）具体访问者

在本问题中，具体访问者是 Student 类和 Teacher 类。Student 类的实例模拟答题；Teacher 类的实例模拟批卷。Student 和 Teacher 类的代码分别如下：

Student. java

```java
import java.util.Scanner;
public class Student implements Visitor {
    String name;
    Student(String name){
        this.name = name;
    }
    Student(){
        name = "匿名";
    }
    public void visit(TestPaper paper){

        Scanner scanner = new Scanner(System.in);
        String question [] = paper.backQuestions();        //试题
        int answerCar [] = paper.backAnswer();             //答题卡
        paper.setName(name);
        System.out.println(name + "开始答题");
        for(int i = 0;i < question.length;i++) {           //开始答题
            System.out.println(question[i]);
            System.out.print("\n请输入解答,回车确认:");
            answerCar[i] = scanner.nextInt();
        }
    }
}
```

Teacher. java

```java
public class Teacher implements Visitor {
    int standardAnswer[] = {18,16,28,3};                   //标准答案
    public void visit(TestPaper paper){
        int totalScore = 0;                                //试卷的总分
        int answerCar [] = paper.backAnswer();             //答题卡
        for(int i = 0;i < answerCar.length;i++) {          //开始批卷
            if(answerCar[i] == standardAnswer[i]){
                totalScore++;
            }
```

```
            }
            System.out.println(paper.name + "试卷的分数" + totalScore);
        }
    }
```

6）应用程序

前面已经使用访问者模式给出了可以使用的类，这些类就是一个小框架，可以使用这个小框架中的类编写应用程序。

在应用程序中，程序首先让具体元素的实例，即试卷，调用抽象元素规定的接收访问者的方法 accept(Visitor visitor)接收具体访问者 Student 类的实例，即模拟学生答卷，然后再让试卷调用抽象元素规定的接收访问者的方法 accept(Visitor visitor)接收具体访问者 Teacher 类的实例，即模拟老师批卷。程序运行效果如图 9.4 所示。

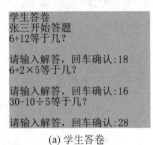

(a) 学生答卷　　　　　　(b) 老师阅卷

图 9.4　程序运行效果

Application.java

```
public class Application{
    public static void main(String args[]) {
        Student stu [] = new Student[3];              //三个学生参加考试，即有三个学生访问试卷
        String name [] = {"张三","李五","赵四"};
        Teacher teacher = new Teacher();              //一个老师批卷
        TestPaper paper[] = new TestPaper[stu.length];  //存放三张试卷
        System.out.println("学生答卷");
        for(int i = 0;i < paper.length;i++) {
            paper[i] = new ArithmeticPaper();
            stu[i] = new Student(name[i]);
            paper[i].accept(stu[i]);
        }
        System.out.println("老师批卷");
        for(int i = 0;i < paper.length;i++) {
            paper[i].accept(teacher);
        }
    }
}
```

第 10 章　状态模式

以下文本框中的内容引自 GoF 所著 *Design Patterns：Elements of Reusable Object Oriented Software* 的中译本及英文版。

> **状态模式（别名：状态对象）**
> 　　允许一个对象在其内部状态改变时改变它的行为，对象看起来似乎修改了它的类。
>
> **State Pattern（Another Name：Objects for States）**
> 　　Allow an object to alter its behavior when its internal state changes. The object will appear to change its class.

以上内容是 GoF 对状态模式的高度概括，结合 10.2.1 节的状态模式的类图可以准确地理解该模式。

10.1　概述

我们知道一个对象调用方法会产生一定的行为效果，但在某些时候，这个方法所产生的行为效果依赖于该对象中的一个称为"状态"的成员变量，即依赖于一个当前对象所组合的成员变量。称该成员变量为对象的状态（也称状态对象）。

例如，一把手枪，调用 firingShooting()方法（开火射击）产生的行为效果依赖于手枪的当前状态。如果手枪的当前状态是有 3 颗子弹并且保险是打开的，那么 firingShooting()方法产生的行为就是射出 1 颗子弹，并引起手枪的当前状态发生变化，即使手枪的当前状态变成有 2 颗子弹并且保险是打开的，即 firingShooting()方法改变了对象的当前状态。在实际问题中，用户的关注点是手枪在某种状态下，firingShooting()方法所产生的效果。

状态模式的本质是一个对象的方法的行为效果依赖于对象的当前状态，当前状态只能是有限种状态中的某一种状态。例如，手枪的当前状态可以是图 10.1 所示的三种状态中的一种："有子弹，保险开""有子弹，保险关"和"无子弹"。在状态模式中，非常关键的一点是对象调用方法可能引起对象的当前状态发生变化，而当前状态的变化又会导致方法的行为发生变化。例如手枪调用 firingShooting()方法，可能导致手枪的状态

图 10.1　手枪的三种状态

改变为"无子弹"，那么手枪再调用 firingShooting()方法的行为将发生变化，即无法射出子弹。正如状态模式定义的那样：对象在其内部状态改变时改变它的行为，对象看起来似乎修改了它的类（实际上没有修改类的代码，见稍后的学习）。

10.2 模式的结构与使用

状态模式包括三种角色。

1. 环境(Context)

环境是一个类,该类含有抽象状态(State)声明的变量,可以引用任何具体状态类的实例。用户对该环境(Context)类的实例在某种状态下的行为感兴趣。

2. 抽象状态(State)

抽象状态是一个接口或抽象类。抽象状态中定义了与环境(Context)的一个特定状态相关的若干个方法,这些方法需依赖环境,以便为环境切换状态。

3. 具体状态(Concrete State)

具体状态是实现(扩展)抽象状态(抽象类)的类。

10.2.1 状态模式的 UML 类图

状态模式的类图如图 10.2 所示。

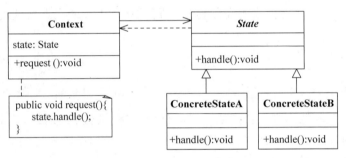

图 10.2 状态模式的类图

10.1 节提到的手枪是状态模式中的环境角色的实例,而"有子弹,保险开""有子弹,保险关"和"无子弹"是具体状态的实例。

10.2.2 结构的描述

下面通过一个简单的问题描述状态模式中所涉及的各个角色。

用状态模式模拟手枪在各种状态下的行为,如射击、装弹等。

1. 环境

环境需要含有抽象状态(State)声明的变量,在本问题中环境是 Gun 类(模拟枪),该类含有一个 State 接口声明的变量 currentState,以便引用具体状态类的实例。Gun 类的代码如下:

Gun.java

```
import java.util.Stack;
public class Gun {
    public Stack < String > clip;                //弹夹(栈是先进后出)
                                                 //当前状态 currentState 可以是下列状态之一:
    public State currentState,
            haveBulletOpeningSafety,             //"有子弹,保险开"状态
            haveBulletClosedSafety,              //"有子弹,保险关"状态
```

```java
            haveNotBullet;                    //"无子弹"状态
        Gun(){
            clip = new Stack<String>();
            clip.push("最后 1 颗子弹。");       //枪里一共装了 3 颗子弹(栈是先进后出)
            for(int k = 1;k <= State.MAXBULLETS - 4;k++){        //MAXBULLETS 的值是 6
                clip.push("1 颗子弹。");
            }
            haveBulletOpeningSafety = new HaveBulletOpeningSafety();
            haveBulletClosedSafety = new HaveBulletClosedSafety();
            haveNotBullet = new HaveNotBullet();
            currentState = haveBulletOpeningSafety;
            //枪的初始状态(也可以指定状态里的其他状态)
        }
        public void firingShooting(){         //开火射击
            currentState = currentState.showGunShoot(this);
        }
        public void openingSafety(){          //打开保险
            currentState = currentState.openGunSafety(this);
        }
        public void closingSafety(){          //关闭保险
            currentState = currentState.closeGunSafety(this);
        }
        public void loadBullet(int amount){
            currentState = currentState.loadBullet(this,amount);
        }
        public int getBulletAmount(){
            return clip.size();
        }
    }
}
```

2. 抽象状态

对于本问题,抽象状态是 State 接口。代码如下:

State.java

```java
public interface State {
    int MAXBULLETS = 6;                                //枪的最大装弹量
    public State showGunShoot(Gun gun);                //显示枪的射击过程(方法依赖环境 Gun)
    public State closeGunSafety(Gun gun);              //关闭枪的保险
    public State openGunSafety(Gun gun);               //打开枪的保险
    public State loadBullet(Gun gun,int amount);       //装弹
}
```

3. 具体状态

对于本问题,共有 3 个具体状态角色,分别是 HaveBulletOpeningSafety、HaveBulletClosedSafety 和 HaveNotBullet 类,分别刻画"有子弹,保险开"状态、"有子弹,保险关"状态和"无子弹"状态。代码如下:

HaveBulletOpeningSafety.java

```java
public class HaveBulletOpeningSafety implements State {       //刻画"有子弹,保险开"状态
    public State showGunShoot(Gun gun){
```

```java
        String bullet = gun.clip.pop();                    //枪的弹夹弹出1颗子弹
        System.out.println("射出" + bullet);
        if(gun.clip.size() >= 1)
            return gun.haveBulletOpeningSafety ;            //返回"有子弹,保险开"状态
        else
            return gun.haveNotBullet;                       //返回"无子弹"状态(保险自动关闭)
    }
    public State closeGunSafety(Gun gun){
        System.out.println("关闭保险。");
        return gun.haveBulletClosedSafety;                  //返回"有子弹,保险关"状态
    }
    public State openGunSafety(Gun gun){
        System.out.println("保险已经打开了,不用再次打开。");
        return this ;                                       //返回当前状态
    }
    public State loadBullet(Gun gun, int amount){
        System.out.println("还有子弹,无法装弹");             //在有子弹的情况下禁止装弹
        return this ;                                       //返回当前状态
    }
}
```

HaveBulletClosedSafety.java

```java
public class HaveBulletClosedSafety implements State {      //刻画"有子弹,保险关"状态
    public State showGunShoot(Gun gun){
        System.out.println("没打开保险,无法射击。");
        return this;                                        //返回当前状态
    }
    public State closeGunSafety(Gun gun){
        System.out.println("保险已经是关闭状态,不用再次关闭。");
        return this;                                        //返回当前状态
    }
    public State openGunSafety(Gun gun){
        System.out.println("打开保险。");
        return gun.haveBulletOpeningSafety ;                //返回"有子弹,保险开"状态
    }
    public State loadBullet(Gun gun, int amount){
        System.out.println("还有子弹,无法装弹");             //在有子弹的情况下禁止装弹
        return this ;                                       //返回当前状态
    }
}
```

HaveNotBullet.java

```java
public class HaveNotBullet implements State {              //刻画"无子弹"状态
    public State showGunShoot(Gun gun){
        System.out.println("没有子弹了,无法射击。");
        return this ;                                       //返回当前状态
    }
    public State closeGunSafety(Gun gun){
        System.out.println("默认就是关闭。");
```

```
            return this ;                           //返回当前状态
        }
        public State openGunSafety(Gun gun){
            System.out.println("默认就是关闭,无法打开。");
            return this ;                           //返回当前状态
        }
        public State loadBullet(Gun gun, int amount){
            //MAXBULLETS 是 State 接口中的常量,值是 6
            amount = amount >= MAXBULLETS?MAXBULLETS - 1:amount;
            System.out.println("装弹" + amount + "颗");
            gun.clip.push("最后 1 颗子弹");            //栈是先进后出
            for(int k = 1;k <= amount - 1;k++){       //一共装了 amount 颗子弹
                gun.clip.push("1 颗子弹");
            }
            return gun.haveBulletClosedSafety ;       //返回"有子弹,保险关"状态
        }
}
```

▶ 10.2.3 模式的使用

前面已经使用状态模式给出了可以使用的类,这些类就是一个小框架,可以使用这个小框架中的类编写应用程序。

下列应用程序(Application.java)使用了状态模式中所涉及的类,演示手枪在射击过程中的状态变化。程序运行效果如图 10.3 所示。

```
射出1颗子弹。
还有子弹,无法装弹
射出1颗子弹。
关闭保险。
保险已经是关闭状态,不用再次关闭。
没打开保险,无法射击。
打开保险。
射出最后1颗子弹。
没有子弹了,无法射击。
没有子弹了,无法射击。
装弹3颗
手枪有3颗子弹
没打开保险,无法射击。
打开保险。
保险已经打开了,不用再次打开。
射出1颗子弹。
手枪当前有2颗子弹。
```

图 10.3   程序运行效果

Application.java

```java
public class Application{
    public static void main(String args[]){
        Gun pistol = new Gun();
        pistol.firingShooting();                //射击
        pistol.loadBullet(5);                   //装弹
        pistol.firingShooting() ;               //射击
        pistol.closingSafety();                 //关闭保险
        pistol.closingSafety();                 //不必连续两次关闭保险
        pistol.firingShooting();                //射击
        pistol.openingSafety();                 //打开保险
```

```
        pistol.firingShooting();           //射击
        pistol.firingShooting();           //射击
        pistol.firingShooting();           //射击
        pistol.firingShooting();           //射击
        pistol.loadBullet(3);              //装弹
        System.out.println("手枪有" + pistol.getBulletAmount() + "颗子弹");
        pistol.firingShooting();           //射击
        pistol.openingSafety();            //打开保险
        pistol.openingSafety();            //不必连续两次打开保险
        pistol.firingShooting();           //射击
        System.out.println("手枪有" + pistol.getBulletAmount() + "颗子弹");
    }
}
```

10.3 状态模式的优点

状态模式具有以下优点：

（1）使用一个类封装对象的一种状态，很容易增加新的状态。

（2）在状态模式中，环境（Context）中不必出现大量的条件判断语句。环境实例所呈现的状态变得更加清晰、容易理解。

（3）使用状态模式可以让用户程序很方便地切换环境实例的状态。

（4）使用状态模式不会让环境的实例中出现内部状态不一致的情况。

扫一扫

视频讲解

10.4 应用举例——咖啡自动售货机

1. 设计要求

我们经常见到一些自动售货机，例如咖啡自动售货机，当投入所要求的钱币（或扫码支付），就会得到一杯咖啡。使用状态模式模拟咖啡自动售货机售卖咖啡。

2. 设计实现

1) 环境

环境需要含有抽象状态声明的变量，本问题中环境是 CoffeeMachine 类（模拟咖啡机），该类含有一个 CoffeeState 接口声明的变量 currentState，以便引用具体状态类的实例。CoffeeState 类的代码如下：

CoffeeState.java

```
import java.util.Stack;
public class CoffeeMachine {
    public Stack<String> coffeeShelf;          //存放咖啡的架(栈是先进后出)
    public CoffeeState currentState,           //当前状态 currentState 可以是下列状态之一:
                haveCoffeeAndCoin,             //"有咖啡,已投币"状态
                haveCoffeeAndNoCoin,           //"有咖啡,未投币"状态
                haveNotCoffee;                 //"无咖啡"状态
    CoffeeMachine(){
        coffeeShelf = new Stack<String>();
```

```java
            coffeeShelf.push("最后1杯咖啡。");          //咖啡机里一共装了3杯(栈是先进后出)
            for(int k = 1;k <= CoffeeState.MAXBCUPS - 98;k++){    //MAXBCUPS的值是100
                coffeeShelf.push("1杯咖啡。");
            }
            haveCoffeeAndCoin = new HaveCoffeeAndCoin();
            haveCoffeeAndNoCoin = new HaveCoffeeAndNoCoin();
            haveNotCoffee = new HaveNotCoffee();
            currentState = haveCoffeeAndNoCoin;         //初始状态(也可以指定状态里的其他状态)
    }
    public void sendCoffee(){                           //咖啡机送出咖啡的方法
        currentState = currentState.popCoffee(this);    //弹出一杯咖啡
    }
    public void coinOperated(){                         //投币
        currentState = currentState.inputCoinToMachine(this);
    }
    public void loadCoffee(int amount){
        currentState = currentState.loadCoffee(this,amount);
    }
    public int getCoffeeAmount(){
        return coffeeShelf.size();
    }
}
```

2) 抽象状态

对于本问题,抽象状态是 CoffeeState 接口。代码如下:

CoffeeState.java

```java
public interface CoffeeState {
    public byte COFFEEPRICE = 5;                        //一杯5元
    public int MAXBCUPS = 100;                          //咖啡机的最大咖啡量(单位:杯)
    public CoffeeState popCoffee(CoffeeMachine machine);    //弹出一杯咖啡
    public CoffeeState inputCoinToMachine(CoffeeMachine machine);    //投币
    public CoffeeState loadCoffee(CoffeeMachine machine, int amount); //补充咖啡
}
```

3) 具体状态

对于本问题,共有 3 个具体状态角色,分别是 HaveCoffeeAndCoin 类、HaveCoffeeAndNoCoin 类和 HaveNotCoffee 类,分别刻画"有咖啡,已投币"状态、"有咖啡,未投币"状态和"无咖啡"状态。代码分别如下:

HaveCoffeeAndCoin.java

```java
public class HaveCoffeeAndCoin implements CoffeeState {    //刻画"有咖啡,已投币"状态
    public CoffeeState popCoffee(CoffeeMachine machine){
        String coffee = machine.coffeeShelf.pop();         //弹出一杯咖啡
        System.out.println("弹出" + coffee);
        if(machine.coffeeShelf.size() >= 1)
            return machine.haveCoffeeAndNoCoin;            //返回"有咖啡,未投币"状态
        else
            return machine.haveNotCoffee;                  //返回"无咖啡"状态
```

```java
    }
    public CoffeeState inputCoinToMachine(CoffeeMachine machine){      //投币
        System.out.println("已经投币,不需要再投币。");
        return this ;                                                   //返回当前状态
    }
    public CoffeeState loadCoffee(CoffeeMachine machine, int amount) {  //补充咖啡
        System.out.println("咖啡机里还有咖啡,不允许装咖啡");
                                                                        //在有咖啡的情况下禁止装咖啡
        return this ;                                                   //返回当前状态
    }
}
```

HaveCoffeeAndNoCoin.java

```java
public class HaveCoffeeAndNoCoin implements CoffeeState {               //刻画"有咖啡,未投币"状态
    public CoffeeState popCoffee(CoffeeMachine machine){
        System.out.println("请投币" + COFFEEPRICE + "元.");              //COFFEEPRICE 是值为 5 的常量
        return this ;                                                   //返回当前状态
    }
    public CoffeeState inputCoinToMachine(CoffeeMachine machine){       //投币
        System.out.print("您投币" + COFFEEPRICE + "元,");
        System.out.println("投币成功,可以让咖啡机送出一杯咖啡。");
        return machine.haveCoffeeAndCoin ;                              //返回"有咖啡,已投币"状态
    }
    public CoffeeState loadCoffee(CoffeeMachine machine, int amount) {  //补充咖啡
        System.out.println("咖啡机里还有咖啡,不允许装咖啡。");            //有咖啡,禁止装咖啡
        return this ;                                                   //返回当前状态
    }
}
```

HaveNotCoffee.java

```java
public class HaveNotCoffee implements CoffeeState {                     //刻画"无咖啡"状态
    public CoffeeState popCoffee(CoffeeMachine machine){
        System.out.println("咖啡机里无咖啡");
        return this ; //返回当前状态
    }
    public CoffeeState inputCoinToMachine(CoffeeMachine machine){       //投币
        System.out.println("咖啡机里没有咖啡了,不能投币。");
        return this ;                                                   //返回当前状态
    }
    public CoffeeState loadCoffee(CoffeeMachine machine, int amount) {  //补充咖啡
        amount = amount >= MAXCUPS?MAXCUPS - 1:amount;
        System.out.println("装咖啡" + amount + "杯。");                   //MAXCUPS 是常量,值是 100
        machine.coffeeShelf.push("最后 1 杯。");                         //栈是先进后出
        for(int k = 1;k <= amount - 1;k++){
            machine.coffeeShelf.push("1 杯咖啡。");                      //一共装了 amount 杯咖啡
        }
        return machine.haveCoffeeAndNoCoin ;                            //返回"有咖啡,未投币"状态
    }
}
```

4）应用程序

前面已经使用状态模式给出了可以使用的类，这些类就是一个小框架，可以使用这个小框架中的类编写应用程序。

下列应用程序（Application.java）使用了状态模式中所涉及的类，演示使用咖啡自动售货机的情况。程序运行效果如图 10.4 所示。

```
请投币5元。
您投币5元,投币成功,可以让咖啡机送出一杯咖啡。
弹出1杯咖啡。
咖啡机当前有2杯咖啡。
您投币5元,投币成功,可以让咖啡机送出一杯咖啡。
弹出1杯咖啡。
您投币5元,投币成功,可以让咖啡机送出一杯咖啡。
弹出最后1杯咖啡。
咖啡机里没有咖啡了,不能投币。
装咖啡66杯。
请投币5元。
您投币5元,投币成功,可以让咖啡机送出一杯咖啡。
弹出1杯咖啡。
咖啡机当前有65杯咖啡。
```

图 10.4　程序运行效果

Application.java

```java
public class Application{
    public static void main(String args[]){
        CoffeeMachine machine = new CoffeeMachine();
        machine.sendCoffee();                //让咖啡机送出咖啡一杯
        machine.coinOperated();              //投币
        machine.sendCoffee();                //让咖啡机送出咖啡一杯
        System.out.println("咖啡机当前有" + machine.getCoffeeAmount() + "杯咖啡。");
        machine.coinOperated();              //投币
        machine.sendCoffee();                //让咖啡机送出咖啡一杯
        machine.coinOperated();              //投币
        machine.sendCoffee();                //让咖啡机送出咖啡一杯
        machine.coinOperated();              //投币
        machine.loadCoffee(66);
        machine.sendCoffee();                //让咖啡机送出咖啡一杯
        machine.coinOperated();              //投币
        machine.sendCoffee();                //让咖啡机送出咖啡一杯
        System.out.println("咖啡机当前有" + machine.getCoffeeAmount() + "杯咖啡。");
    }
}
```

第 11 章　装饰模式

以下文本框中的内容引自 GoF 所著 *Design Patterns*: *Elements of Reusable Object Oriented Software* 的中译本及英文版。

> **装饰模式（别名：包装器）**
> 　　动态地给对象添加一些额外的职责。就功能来说，装饰模式相比生成子类更为灵活。
> **Decorator Pattern（Another Name：Wrapper）**
> 　　Attach additional responsibilities to an object dynamically. Decorators provide a flexible alternative to subclassing for extending functionality.

以上内容是 GoF 对装饰模式的高度概括，结合 11.2.1 节的装饰模式的类图可以准确地理解该模式。

11.1　概述

在许多设计中，可能需要改进类的某个对象的功能，而不是该类创建的全部对象。例如，麻雀类的实例（麻雀）能连续飞行 100 米，如果用麻雀类创建了 5 只麻雀，那么这 5 只麻雀都能连续飞行 100 米。假如想让其中一只麻雀，例如名字是 blackSparrow（如图 11.1(a)所示）的麻雀，连续飞行 150 米，那应当怎样做呢？我们不想通过修改麻雀类的代码（也可能根本不允许修改）使得麻雀类创建的麻雀都能连续飞行 150 米，这也不符合我们的初衷：改进类的某个对象的功能。

(a) blackSparrow能连续飞行100米　　(b) blackSparrowDecorator能连续飞行150米

图 11.1　麻雀连续飞行

一种比较好的办法是给 blackSparrow 装上电子翅膀（对 blackSparrow 进行装饰）。电子翅膀可以使 blackSparrow 不使用自己的翅膀就能飞行 50 米，那么 blackSparrow 就能连续飞行 150 米，因为 blackSparrow 首先使用自己的翅膀飞行 100 米，然后电子翅膀开始工作再飞行 50 米。例如，图 11.1(b)中的"麻雀的装饰类"创建的对象（给被装饰者 blackSparrow 装上电子翅膀后的对象）blackSparrowDecorator，就能连续飞行 150 米。

装饰模式是动态地扩展一个对象的功能，而不需要改变原始类代码的一种成熟模式。

11.2 模式的结构与使用

装饰模式的结构中包括四种角色。

1. 抽象组件（Component）

抽象组件是一个抽象类。抽象组件定义了"被装饰者"需要被"装饰"的方法。

2. 具体组件（ConcreteComponent）

具体组件是抽象组件的一个子类。具体组件的实例称作被装饰者。

3. 装饰（Decorator）

装饰也是抽象组件的一个子类，称作具体组件的实例的装饰类，简称装饰类。但装饰类还包含一个抽象组件声明的对象。装饰类可以是抽象类，也可以是非抽象类，如果是非抽象类，那么该类的实例称作组件的实例的装饰者，即"被装饰者"的装饰者，简称装饰者。

4. 具体装饰（ConcreteDecotator）

具体装饰是装饰的一个非抽象子类，是具体组件的实例的装饰类。具体装饰的实例称作装饰者。

11.2.1 装饰模式的 UML 类图

装饰模式的类图如图 11.2 所示。

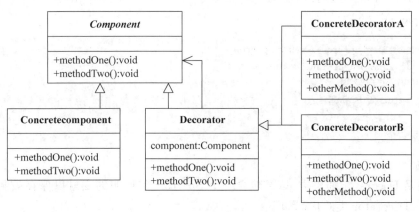

图 11.2 装饰模式的类图

11.1 节提到的名字是 blackSparrow 的麻雀（见图 11.1(a)）是具体组件的实例，即一只能连续飞行 100 米的麻雀，而 blackSparrowDecorator 是它的装饰者。由 UML 图可知，装饰类也是抽象组件的子类，因此也可以把 blackSparrowDecorator 当作被装饰者，使用装饰类对其进行装饰（给 blackSparrowDecorator 增加电子翅膀）再得到一个装饰者，即名字是 blackSparrowDecoratorAgain 的麻雀，那么 blackSparrowDecoratorAgain 就能连续飞行 200 米，如图 11.3 所示。

图 11.3 blackSparrowDecoratorAgain 能连续飞行 200 米

11.2.2 结构的描述

下面通过一个简单的问题来描述装饰模式中所涉及的各个角色。

假设系统中有一个 Sparrow 抽象类以及它的一个 BlackSparrow 子类。BlackSparrow 子类重写了 Sparrow 抽象类的 fly() 方法，使得 BlackSparrow 类创建的麻雀调用 fly() 方法能连续飞行 100 米。现在，用户需要两只麻雀，分别能连续飞行 150 米和 200 米。不允许修改 BlackSparrow 类。请使用装饰模式设计类，使得系统创建出能连续飞行 150 米和 200 米的麻雀。

1. 抽象组件

抽象组件是 Sparrow 抽象类，该类定义了被装饰者，即麻雀需要被装饰的 fly() 方法。Sparrow 类的代码如下：

Sparrow.java

```java
public abstract class Sparrow {
    public abstract int fly();
}
```

2. 具体组件

具体组件是 BlackSparrow 类，该类在重写 Sparrow 类的 fly() 方法时，将 fly() 方法的返回值设置为 100。BlackSparrow 类的代码如下：

BlackSparrow.java

```java
public class BlackSparrow extends Sparrow{
    public final int DISTANCE = 100;
    public int fly() {
        return DISTANCE;
    }
}
```

3. 装饰

装饰类是 Decorator 抽象类，该类包含一个 Sparrow 类声明的变量以保存被装饰者（即麻雀）的引用。Decorator 类的代码如下：

Decorator.java

```java
public abstract class Decorator extends Sparrow{
    Sparrow sparrow;
    public Decorator(){
    }
    public Decorator(Sparrow sparrow){
        this.sparrow = sparrow;
    }
}
```

4. 具体装饰

具体装饰是 SparrowDecorator 类。SparrowDecorator 类的代码如下：

第 11 章　装饰模式

SparrowDecorator.java

```java
public class SparrowDecorator extends Decorator{
    public final int DISTANCE = 50;           //下面的 eleFly()方法能飞 DISTANCE 米
    SparrowDecorator(Sparrow sparrow){
        super(sparrow);
    }
    public int fly(){
        int distance = 0;
        //装饰者先让被装饰者 sparrow 调用 fly(),然后装饰者再调用 eleFly()
        distance = sparrow.fly() + eleFly();
        return distance;
    }
    private int eleFly(){                     //装饰者新添加的方法
        //客户程序只有调用 fly 方法才可以使用 eleFly 方法
        return DISTANCE;
    }
}
```

11.2.3　模式的使用

前面已经使用装饰模式给出了可以使用的类,这些类就是一个小框架,可以使用这个小框架中的类编写应用程序。

下列应用程序(Application.java)使用了装饰模式中所涉及的类,演示两只分别能连续飞行 150 米和 200 米的麻雀。程序运行效果如图 11.4 所示。

没被装饰的麻雀sparrow能飞行100
被装饰的麻雀sparrow能飞150
再次被装饰的麻雀sparrow能飞行200

图 11.4　程序运行效果

Application.java

```java
public class Application{
    public static void main(String args[]) {
        Sparrow sparrow = new BlackSparrow();
        System.out.println("没被装饰的麻雀 sparrow 能飞行" + sparrow.fly());
        //得到 sparrow 的装饰者(将装饰者的引用继续赋值给 sparrow 对象)
        sparrow = new SparrowDecorator(sparrow);
        System.out.println("被装饰的麻雀 sparrow 能飞行" + sparrow.fly());
        sparrow = new SparrowDecorator(sparrow);
        System.out.println("再次被装饰的麻雀 sparrow 能飞行" + sparrow.fly());
    }
}
```

注意:通过继承也可以改进对象的行为,对于某些简单的问题这样做未尝不可,但是如果考虑到系统的扩展性,就应当注意面向对象的一个基本原则:少用继承,多用组合。如果使用继承,但不采用装饰模式,为了满足用户的需求,就需要不断地增加类似 BlackSparrow 的各种 Sparrow 的子类,导致系统中子类数目膨胀,降低系统的可维护性。

11.3　装饰模式的优点

装饰模式具有以下优点:
(1) 使用装饰模式,程序可以动态地增强类的某个对象的功能,而又不影响该类的其他对象。

(2) 被装饰者和装饰者是松耦合关系。由于装饰（Decorator）仅仅依赖于抽象组件（Component），因此具体装饰只知道它要装饰的对象是抽象组件某一个子类的实例，但不需要知道是哪一个具体子类。

(3) 装饰模式满足"开-闭"原则。不必修改具体组件，就可以增加新的针对该具体组件的具体装饰。

(4) 可以使用多个具体装饰来装饰具体组件的实例。

扫一扫

视频讲解

11.4 应用举例——读取单词表

1. 设计要求

当前系统已有一个抽象类 ReadWord，该类有一个抽象方法 readWord()，另外，系统还有一个 ReadWord 类的子类 ReadEnglishWord，该类的 readWord() 方法可以读取一个由英文单词构成的文本文件 word.txt。word.txt 的格式是每行只有一个单词，例如 word.txt 的前三行内容如下：

```
arrange
correct
intelligence
```

系统已有类的类图如图 11.5 所示。

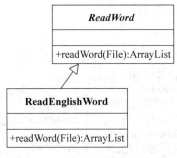

图 11.5 系统已有类的类图

ReadWord 和 ReadEnglishWord 类的代码如下：

ReadWord.java

```java
import java.io.File;
import java.util.ArrayList;
public abstract class ReadWord {
    public abstract ArrayList<String> readWord(File file);
}
```

ReadEnglishWord.java

```java
import java.io.*;
import java.util.ArrayList;
public class ReadEnglishWord extends ReadWord {
    public ArrayList<String> readWord(File file){
```

```java
        ArrayList<String> list = new ArrayList<String>();
        try{
            FileReader inOne = new FileReader(file);
            BufferedReader inTwo = new BufferedReader(inOne);
            String s = null;
            while((s = inTwo.readLine())!= null){
                list.add(s);
            }
            inTwo.close();
            inOne.close();
        }
        catch(IOException exp){
            System.out.println(exp);
        }
        return list;
    }
}
```

一些用户在使用该系统时，只使用 ReadWord 类的对象调用 readWord() 方法读取文件中的英文单词，即这部分用户只想要英文单词。

现在有部分用户要求不仅要得到系统提供的英文单词，而且要得到该单词的汉语解释，还有一些用户要求不仅要得到系统提供的英文单词以及解释，还要得到英文单词的例句。要求不修改现有系统的代码以及 word.txt 文件，使用装饰模式对系统进行扩展，以满足用户的需求。

2. 设计实现

1) 抽象组件

抽象组件是原系统中已有的 ReadWord 类，见前面"设计要求"中的 ReadWord.java。

2) 具体组件

具体组件是原系统中已有的 ReadEnglishWord 类，见前面"设计要求"中的 ReadEnglishWord.java。

3) 装饰

在原系统中添加一个装饰，该装饰是 Decorator 类，代码如下：

Decorator.java

```java
public abstract class Decorator extends ReadWord {
    ReadWord reader;
    public Decorator(){
    }
    public Decorator(ReadWord reader){
        this.reader = reader;
    }
}
```

4) 具体装饰与相关文件

具体装饰是 WordDecorator 类（Decorator 类的子类），该类使用文本文件 chinese.txt 或 englishSentence.txt 来装饰"被装饰者"。chinese.txt 和 englishSentence.txt 中每行的文本内容分别是原系统 word.txt 中对应行上单词的汉语解释和英语例句。

chinese.txt 的前三行内容如下：

vt. 安排,排列,整理。
adj. 正确的,合适的,恰当的。v. 改正,批改。
n. 智力,天分,智慧,理解力。

englishSentence.txt 的前三行内容如下:

Have you arranged to meet him?

I think you've made the correct decision.

The question was an insult to our intelligence.

WordDecorator 类的代码如下:

WordDecorator.java

```java
import java.io.*;
import java.util.ArrayList;
public class WordDecorator extends Decorator{
    File decoratorFile;
    WordDecorator(ReadWord reader,File decoratorFile){
        super(reader);
        this.decoratorFile = decoratorFile;
    }
    public ArrayList<String> readWord(File file){
        ArrayList<String> list = reader.readWord(file);
        ArrayList<String> backList = new ArrayList<String>();
        try{
            FileReader inOne = new FileReader(decoratorFile);
            BufferedReader inTwo = new BufferedReader(inOne);
            String decorativeMaterials = null;
            for(int m = 0;m < list.size();m++) {
                String word = list.get(m);
                decorativeMaterials = inTwo.readLine();
                //用 decorativeMaterials 装饰单词
                word = word.concat(" | " + decorativeMaterials);
                backList.add(word);
            }
            inTwo.close();
            inOne.close();
        }
        catch(IOException exp){
            System.out.println(exp);
        }
        return backList;
    }
}
```

5)应用程序

前面已经使用装饰模式给出了可以使用的类,这些类就是一个小框架,可以使用这个小框架中的类编写应用程序。

下列应用程序(Application.java)使用了装饰模式中所涉及的类,输出带有汉语解释的英文单词,也输出既带有汉语解释又带有英文例句的英文单词。程序运行效果如图 11.6 所示。

```
英文单词：
arrange
correct
intelligence
英文单词、中文解释：
arrange     | vt.安排，排列，整理。
correct     | adj.正确的，合适的，恰当的。v.改正，批改。
intelligence | n.智力，天分，智慧，理解力，
英文单词、中文解释和英文例句：
arrange     | vt.安排，排列，整理。      | Have you arranged to meet him?
correct     | adj.正确的，合适的，恰当的。v.改正，批改。  | I think you've made the correct decision.
intelligence | n.智力，天分，智慧，理解力， | The quesion was an insult to our intelligence.
```

图 11.6　程序运行效果

Application.java

```java
import java.util.ArrayList;
import java.io.File;
public class Application{
    public static void main(String args[]) {
        ArrayList<String> wordList = new ArrayList<String>();
        ReadWord reader = new ReadEnglishWord();
        wordList = reader.readWord(new File("word.txt"));
        System.out.println("英文单词：");
        for(int i = 0;i < wordList.size();i++){
            System.out.println(wordList.get(i));
        }
        reader = new WordDecorator(reader,new File("chinese.txt"));
        wordList = reader.readWord(new File("word.txt"));
        System.out.println("英文单词、中文解释：");
        for(int i = 0;i < wordList.size();i++){
            System.out.println(wordList.get(i));
        }
        reader = new WordDecorator(reader,new File("englishSentence.txt"));
        wordList = reader.readWord(new File("word.txt"));
        System.out.println("英文单词、中文解释和英文例句：");
        for(int i = 0;i < wordList.size();i++){
            System.out.println(wordList.get(i));
        }
    }
}
```

第 12 章　生成器模式

以下文本框中的内容引自 GoF 所著 *Design Patterns*：*Elements of Reusable Object Oriented Software* 的中译本及英文版。

> **生成器模式**
> 　　将一个复杂对象的构建与它的表示分离，使同样的构建过程可以创建不同的表示。
>
> **Builder Pattern**
> 　　Separate the construction of a complex object from its representation so that the same construction process can create different representations.

以上内容是 GoF 对生成器模式的高度概括，结合 12.2.1 节的生成器模式的类图可以准确地理解该模式。

12.1　概述

在编写一个类的构造方法中的代码时，可能需要更大的灵活性才能满足用户对该类所创建的对象的需求。例如，为了创建有三块月饼的月饼盒，编写了 Box 类，代码如下：

```
public class Box {//含有三块月饼的盒
    String moonCakeOne,
           moonCakeTwo,
           moonCakeThree;
    Box() {
        moonCakeOne = "五仁月饼";
        moonCakeTwo = "豆沙月饼";
        moonCakeThree = "黑芝麻月饼";
    }
}
```

观察上述代码，不难发现，如果用 Box 类创建三个月饼盒，那么这些盒子里面的三块月饼都分别是"五仁月饼""豆沙月饼"和"黑芝麻月饼"，如图 12.1 所示。

图 12.1　三个盒子里面的月饼品种相同

有些用户对月饼盒有特殊的要求，例如希望月饼盒里都是"五仁月饼"，或者一块"五仁月饼"、两块"黑芝麻月饼"等。显然上面的 Box 类无法满足这样的需求，因为 Box 类的构造方法中的编码属于"硬编码"(没有任何弹性的编码),无法满足需求的变化(如果用户的需求没有变化,那么上面的Box类的代码就没有问题)。按照面向抽象编程的原则,不应该在Box类的构造方法中进行硬编码,而是应该将Box类的对象的构造过程的任务交给另外一个类的对象去完成,即生成器。在生成器模式中,封装一个复杂对象构造过程的类称作生成器。

12.2 模式的结构与使用

生成器模式的结构中包括四种角色。

1. 产品(Product)

产品角色是一个类,但该类的对象的构造过程比较复杂。这个类的对象在本模式中称作"复杂对象"。

2. 抽象生成器(Builder)

抽象生成器角色是一个接口,该接口定义了若干个方法,每个方法都是构造复杂对象(产品(Product)的实例)过程中的一个步骤。另外,该接口还必须定义一个返回复杂对象的方法。

3. 具体生成器(ConcreteBuilder)

具体生成器角色是实现Builder接口的类,具体生成器将实现Builder接口所定义的方法。

4. 指挥者(Director)

指挥者是生成器模式中非常重要的角色,指挥者是一个类,该类的对象负责向用户提供复杂对象(产品(Product)的实例)。在生成器模式中,指挥者始终按照抽象生成器给出的构建复杂对象的步骤来指挥具体生成器去构造复杂对象,并返回这个复杂对象给用户。用户需要请求指挥者返回给自己一个复杂对象,而不是请求具体生成器。指挥者类面向抽象生成器,为用户屏蔽掉了具体生成器构造复杂对象的细节,有利于代码的维护和系统的扩展。

▶ 12.2.1 生成器模式的 UML 类图

生成器模式的类图如图12.2所示。

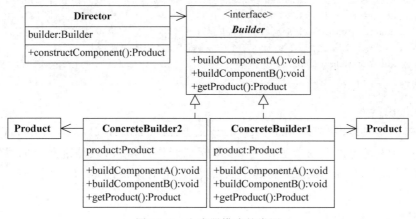

图 12.2 生成器模式的类图

从UML图可以看出,如果系统再增加具体生成器,不会引起指挥者代码的修改。具体生成器完成复杂对象的内部的细节,即每个具体生成器处理复杂对象(Product类的实例)的细节不尽相同。指挥者会在某个方法，例如 UML 图中给出的 constructComponent（）方法中，使用抽象生成器给出的那些方法，即始终按照抽象生成器给出的创建复杂对象的步骤（模式定义中提到的"同样的构建过程"）来指挥具体生成器去构造复杂对象，然后返回这个复杂对象给用户。这体现了生成器模式的定义："将一个复杂对象的构建（指挥者）与它的表示（具体生成器）分离，使同样的构建过程可以创建不同的表示。"

12.2.2 结构的描述

下面通过一个简单的问题来描述生成器模式中所涉及的各个角色。

月饼盒里需要有不同种类的月饼，使用生成器模拟月饼盒。

1. 产品

产品角色是 Box 类，其代码如下：

Box.java

```java
public class Box {                              //含有三块月饼的盒
    String moonCakeOne,
           moonCakeTwo,
           moonCakeThree;
    public void showMoonCakeInBox(){
        System.out.println("月饼盒里的月饼:");
        System.out.println("月饼 1:" + moonCakeOne);
        System.out.println("月饼 2:" + moonCakeTwo);
        System.out.println("月饼 3:" + moonCakeThree);
    }
}
```

2. 抽象生成器

抽象生成器是 Builder 接口，其代码如下：

Builder.java

```java
public interface Builder {
    public abstract void buildWurenMoonCake();       //模拟制作五仁月饼
    public abstract void buildBeanMoonCake();        //模拟制作豆沙月饼
    public abstract void buildSesameMoonCake();      //模拟制作黑芝麻月饼
    //返回一个 Box 对象之前，需要上述三个方法（即步骤）都能执行成功
    public abstract Box getBox();                    //返回一个 Box 类的实例
}
```

3. 具体生成器

有三个具体生成器，分别是 BuildTypeOneBox 类、BuildTypeTwoBox 类和 BuildTypeThreeBox 类。三个类的实例都返回 Box 类型的对象，即模拟制作月饼盒，但月饼盒里的月饼种类不尽相同。例如，用 BuildTypeOneBox 类的实例，月饼盒里的月饼都是"五仁月饼"。三个类的代码如下：

BuildTypeOneBox.java

```java
public class BuildTypeOneBox implements Builder {    //月饼盒中的三块月饼都是五仁月饼
    Box box;
    public BuildTypeOneBox(){
```

```
            box = new Box();
        }
        public void buildWurenMoonCake(){              //制作五仁月饼
            System.out.println("用杏仁、桃仁、花生仁、麻仁和瓜子仁制作的月饼");
            box.moonCakeOne = "五仁月饼";
            box.moonCakeTwo = "五仁月饼";
            box.moonCakeThree = "五仁月饼";
        }
        public void buildBeanMoonCake(){               //没有制作豆沙月饼
        }
        public void buildSesameMoonCake(){             //没有制作黑芝麻月饼
        }
        public Box getBox() {
            return box;
        }
    }
```

BuildTypeTwoBox.java

```
    public class BuildTypeTwoBox implements Builder {      //月饼盒中有一块五仁月饼、两块豆沙月饼
        Box box;
        public BuildTypeTwoBox(){
            box = new Box();
        }
        public void buildWurenMoonCake(){              //制作五仁月饼
            System.out.println("用杏仁、桃仁、花生仁、麻仁和瓜子仁制作的月饼");
            box.moonCakeOne = "五仁月饼";
        }
        public void buildBeanMoonCake(){               //制作豆沙月饼
            System.out.println("用红豆沙制作的月饼");
            box.moonCakeTwo = "豆沙月饼";
            box.moonCakeThree = "豆沙月饼";
        }
        public void buildSesameMoonCake(){             //没有制作黑芝麻月饼
        }
        public Box getBox() {
            return box;
        }
    }
```

BuildTypeThreeBox.java

```
    public class BuildTypeThreeBox implements Builder {
                                        //月饼盒中五仁月饼、豆沙月饼、黑芝麻月饼各一块
        Box box;
        public BuildTypeThreeBox(){
            box = new Box();
        }
        public void buildWurenMoonCake(){              //制作五仁月饼
            System.out.println("用杏仁、桃仁、花生仁、麻仁和瓜子仁制作的月饼");
            box.moonCakeOne = "五仁月饼";
        }
```

```java
    public void buildBeanMoonCake(){           //制作豆沙月饼
        System.out.println("用红豆沙制作的月饼");
        box.moonCakeTwo = "豆沙月饼";
    }
    public void buildSesameMoonCake(){         //制作黑芝麻月饼
        System.out.println("用黑芝麻制作的月饼");
        box.moonCakeThree = "黑芝麻月饼";
    }
    public Box getBox() {
        return box;
    }
}
```

4. 指挥者

指挥者是 Director 类,该类中的 manufacturingSteps()方法按照抽象生成器接口给出的方法步骤指挥具体生成器返回 Box 类的实例并将该实例返回给用户。这一点是生成器模式中非常重要的环节,即各个具体生成器听从统一的指挥,最终体现了生成器模式的定义:将一个复杂对象的构建与它的表示分离,使同样的构建过程可以创建不同的表示。Director 类的代码如下:

Director.java

```java
public class Director{
    private Builder builder;
    Director(Builder builder){
        this.builder = builder;
    }
    public Box manufacturingSteps(){           //制作月饼盒的步骤(即构建产品实例的步骤)
        builder.buildWurenMoonCake();          //制作五仁月饼
        builder.buildBeanMoonCake();           //制作豆沙月饼
        builder.buildSesameMoonCake();         //制作黑芝麻月饼
        Box box = builder.getBox();            //返回月饼盒
        return box;
    }
}
```

▶ 12.2.3 模式的使用

前面已经使用生成器模式给出了可以使用的类,这些类就是一个小框架,可以使用这个小框架中的类编写应用程序。

应用程序将使用指挥角色创建一个"指挥者"对象,并将一个具体生成器传递给指挥者,指挥者请求具体生成器开始构造用户所需要的 Box 类型的对象(月饼盒),如果该具体生成器成功地构造出 Box 类型的对象,指挥者就返回这个对象。下列应用程序(Application.java)使用了生成器模式中所涉及的类,输出三个月饼盒里的月饼种类。程序运行效果如图 12.3 所示。

Application.java

```java
public class Application{
    public static void main(String args[]){
```

```
            用杏仁、桃仁、花生仁、麻仁和瓜子仁制作的月饼
            月饼盒里的月饼：
            月饼1：五仁月饼
            月饼2：五仁月饼
            月饼3：五仁月饼
            用杏仁、桃仁、花生仁、麻仁和瓜子仁制作的月饼
            用红豆沙制作的月饼
            月饼盒里的月饼：
            月饼1：五仁月饼
            月饼2：豆沙月饼
            月饼3：豆沙月饼
            用杏仁、桃仁、花生仁、麻仁和瓜子仁制作的月饼
            用红豆沙制作的月饼
            用黑芝麻制作的月饼
            月饼盒里的月饼：
            月饼1：五仁月饼
            月饼2：豆沙月饼
            月饼3：黑芝麻月饼
```

图 12.3　程序运行效果

```
        Builder build = new BuildTypeOneBox();
        Director director = new Director(build);
        Box box = director.manufacturingSteps();
        box.showMoonCakeInBox();
        director = new Director(new BuildTypeTwoBox());
        box = director.manufacturingSteps();
        box.showMoonCakeInBox();
        director = new Director(new BuildTypeThreeBox());
        box = director.manufacturingSteps();
        box.showMoonCakeInBox();
    }
}
```

注意：用户程序不要直接请求具体生成器返回 Box 类型的对象，如果将 main 方法中的代码

　　Box box = director.manufacturingSteps();

修改为：

　　Box box = build.getBox();

那么

　　box.showMoonCakeInBox();

代码输出的结果就变成：

　　月饼盒里的月饼：
　　月饼 1：null
　　月饼 2：null
　　月饼 3：null

原因是没有按照指挥者的要求(即 manufacturingSteps()方法)生产月饼。所以月饼盒里没有月饼。

12.3　生成器模式的优点

生成器模式具有以下优点：

(1) 生成器模式将对象的构造过程封装在具体生成器中，指挥者使用不同的具体生成器

就可以得到该对象的不同表示。

(2) 生成器模式将对象的构造过程从创建该对象的类中分离出来,使用户无须了解该对象的具体组件。

(3) 可以更加精细有效地控制对象的构造过程。生成器将对象的构造过程分解成若干步骤,这就使程序可以更加精细、有效地控制整个对象的构造。

(4) 生成器模式将对象的构造过程与创建该对象类解耦,使对象的创建更加灵活、有弹性。

(5) 当增加新的具体生成器时,不必修改指挥者的代码,即该模式满足"开-闭"原则。

扫一扫

视频讲解

12.4 应用举例——日历牌

1. 设计要求

我们知道,中国式的日历牌的每个星期的第一天是星期一,最后一天是星期日;欧美式的日历牌的每个星期的第一天是星期日,最后一天是星期六。要求使用生成器模式为用户提供中国式和欧美式的日历牌。

2. 设计实现

1) 产品

产品角色是 CalendarCard 类。代码如下。

CalendarCard.java

```
import javax.swing.*;
public class CalendarCard {                    //日历牌
    public String title;                       //日历牌的标题
    public String [] weekTitle;                //日历牌的星期标题
    public String [][] daysOfMonth;            //用来存放一个月中号码的数组
    public int year,                           //日历上的年
        month = 1;                             //日历上的月
    public void showCalendarCard(){
        JTable table;
        table = new JTable(daysOfMonth,weekTitle);
        table.setFont(new java.awt.Font("",java.awt.Font.BOLD,30));
        table.setRowHeight(60);
        JDialog dialog = new JDialog();
        dialog.setTitle(title);
        dialog.add(new JScrollPane(table));
        dialog.setBounds(10,60,500,460);
        dialog.setVisible(true);
        dialog.setDefaultCloseOperation(JFrame.DISPOSE_ON_CLOSE);
    }
}
```

2) 抽象生成器

抽象生成器是 Builder 接口。代码如下:

Builder.java

```
public interface Builder {
    public abstract void buildYearAndMonth(int y,int m);
```

```java
    public abstract void buildTitle();
    public abstract void buildWeekTitle();
    public abstract void buildDaysOfMonth();
    //以上方法是构建步骤
    public abstract CalendarCard getCalendarCard();
}
```

3）具体生成器

有两个具体生成器：ChineseCalendarCardBuilder 类和 AmericanCalendarCardBuilder 类。代码分别如下：

ChineseCalendarCardBuilder.java

```java
import java.time.LocalDate;
import java.time.DayOfWeek;
public class ChineseCalendarCardBuilder implements Builder {    //负责构造中国式的日历
    CalendarCard calendarCard ;
    public ChineseCalendarCardBuilder(){
        calendarCard = new CalendarCard();
    }
    public void buildYearAndMonth(int y,int m){
        calendarCard.year = y;
        calendarCard.month = m;
    }
    public void buildTitle(){
        calendarCard.title =
        new String("中国式日历牌: " + calendarCard.year + "/" + calendarCard.month);
    }
    public void buildWeekTitle(){
        String weekHabitMess[] = {"一","二","三","四","五","六","日"};
        calendarCard.weekTitle = weekHabitMess;
    }
    public void buildDaysOfMonth(){
        LocalDate date = LocalDate.of(calendarCard.year,calendarCard.month,1);    //日期是1
        int daysAmount = date.lengthOfMonth();              //得到该月有几天
        int weekSerial = getSerial(date.getDayOfWeek());    //得到日期1号的星期顺序
        int a[] = new int[daysAmount + weekSerial];
        for(int i = 0;i < weekSerial - 1;i++){
            a[i] = 0;
        }
        //把全部日期号码放到一维数组 a 中
        for(int i = weekSerial - 1,number = 1;i < a.length - 1;i++){
            a[i] = number;
            number++;
        }
        calendarCard.daysOfMonth = new String[6][7];        //需要6行7列的数组来存放号码
        for(int i = 0,m = 0;i < 6;i++) {
            for(int j = 0;j < 7;j++) {
                if(m < a.length) {
                    if(a[m] == 0)
                        calendarCard.daysOfMonth[i][j] = "";
```

```java
                    else
                        calendarCard.daysOfMonth[i][j] = "" + a[m];
                }
                else {
                    calendarCard.daysOfMonth[i][j] = "";
                }
                m++;
            }
        }
    }
    public CalendarCard getCalendarCard(){
        return calendarCard;
    }
    private int getSerial(DayOfWeek x) {
        int n = 0;
        switch(x) {
          case SUNDAY: n = 7;       //SUNDAY 是枚举 DayOfWeek 中的常量
                      break;
          case MONDAY: n = 1;
                      break;
          case TUESDAY: n = 2;
                      break;
          case WEDNESDAY:n = 3;
                      break;
          case THURSDAY: n = 4;
                      break;
          case FRIDAY: n = 5;
                      break;
          case SATURDAY: n = 6;
                      break;
        }
        return n;
    }
}
```

AmericanCalendarCardBuilder.java

```java
import java.time.LocalDate;
import java.time.DayOfWeek;
public class AmericanCalendarCardBuilder implements Builder {    //负责构造欧美式的日历
    CalendarCard calendarCard ;
    public AmericanCalendarCardBuilder(){
        calendarCard = new CalendarCard();
    }
    public void buildYearAndMonth(int y, int m){
        calendarCard.year = y;
        calendarCard.month = m;
    }
    public void buildTitle(){
        calendarCard.title =
   new String("EuropeanAmericanStyle:" + calendarCard.year + "/" + calendarCard.month);
```

```java
    }
    public void buildWeekTitle(){
        String weekHabitMess[] = {"Sun"," Mon"," Tue"," Wed"," Thu"," Fri"," Sat"};
        calendarCard.weekTitle = weekHabitMess;
    }
    public void buildDaysOfMonth(){
        LocalDate date = LocalDate.of(calendarCard.year,calendarCard.month,1);    //日期1号
        int daysAmount = date.lengthOfMonth();                  //得到该月有几天
        int weekSerial = getSerial(date.getDayOfWeek());        //得到日期1号的星期顺序
        //日期号码放到一维数组 a 中(注意和中国式的不同)
        int a[] = new int[daysAmount + weekSerial];
        for(int i = 0;i < weekSerial;i++){                      //注意和中国式的不同
            a[i] = 0;
        }
        //把全部日期号码放到一维数组 a 中
        for(int i = weekSerial,number = 1;i < a.length;i++){
            a[i] = number;
            number++;
        }
        calendarCard.daysOfMonth = new String[6][7];        //需要6行7列的数组来存放号码
        for(int i = 0,m = 0;i < 6;i++) {
            for(int j = 0;j < 7;j++) {
                if(m < a.length) {
                    if(a[m] == 0)
                        calendarCard.daysOfMonth[i][j] = "";
                    else
                        calendarCard.daysOfMonth[i][j] = "" + a[m];
                }
                else {
                    calendarCard.daysOfMonth[i][j] = "";
                }
                m++;
            }
        }
    }
    public CalendarCard getCalendarCard(){
        return calendarCard;
    }
    private int getSerial(DayOfWeek x) {
        int n = 0;
        switch(x) {
            case SUNDAY: n = 0;                 //SUNDAY 是枚举 DayOfWeek 中的常量(与中国式日历牌不同)
                        break;
            case MONDAY: n = 1;
                        break;
            case TUESDAY: n = 2;
                        break;
            case WEDNESDAY:n = 3;
                        break;
            case THURSDAY: n = 4;
                        break;
            case FRIDAY: n = 5;
                        break;
            case SATURDAY: n = 6;
                        break;
```

```
            }
            return n;
        }
    }
```

4）指挥者

指挥者是 Director 类。代码如下：

Director.java

```
public class Director{
    private Builder builder;
    int year,month;
    Director(Builder builder,int year,int month){
        this.builder = builder;
        this.year = year;
        this.month = month;
    }
    public CalendarCard constructCalendarCard(){
        builder.buildYearAndMonth(year,month);
        builder.buildTitle();
        builder.buildWeekTitle();
        builder.buildDaysOfMonth();
        //以上方法是构建步骤
        return builder.getCalendarCard();
    }
}
```

5）应用程序

前面已经使用生成器模式给出了可以使用的类，这些类就是一个小框架，可以使用这个小框架中的类编写应用程序。

应用程序将使用模式中的指挥角色创建一个"指挥者"对象，并将一个具体生成器传递给指挥者，指挥者请求具体生成器开始构造用户所需要的 CalendarCard 类的对象，如果该具体生成器成功地构造出 CalendarCard 类的对象，指挥者就返回这个对象。下列应用程序（Application.java）显示了中国式日历牌和欧美式日历牌。程序运行效果如图 12.4 所示。

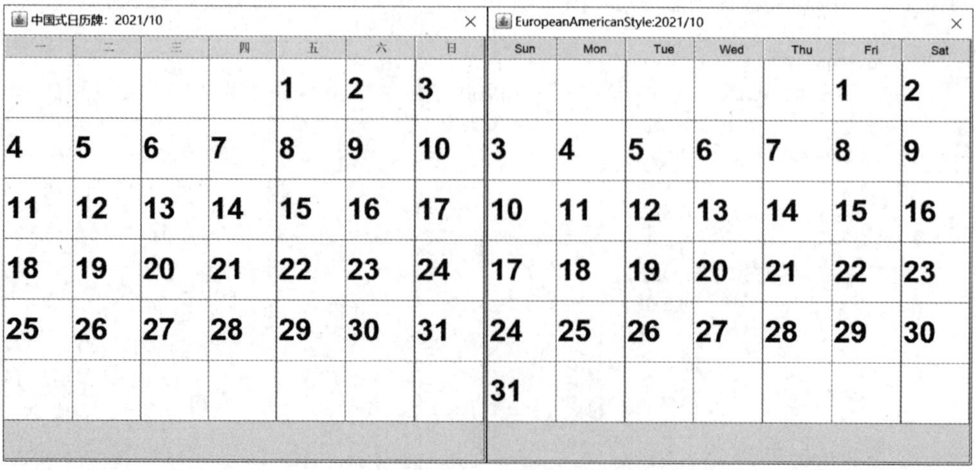

图 12.4　程序运行效果

Application.java

```java
import java.util.Scanner;
public class Application {
    public static void main(String args[]){
        Scanner scanner = new Scanner(System.in);
        System.out.print("输入年和月(用空格分隔),回车确认:");
        int year = scanner.nextInt();
        int month = scanner.nextInt();
        Builder builder = new ChineseCalendarCardBuilder();
        Director director = new Director(builder,year,month);
        CalendarCard cardChina = director.constructCalendarCard();
        cardChina.showCalendarCard();
        builder = new AmericanCalendarCardBuilder();
        director = new Director(builder,year,month);
        CalendarCard cardAmerican = director.constructCalendarCard();
        cardAmerican.showCalendarCard();
    }
}
```

第 13 章 工厂方法模式

以下文本框中的内容引自 GoF 所著 *Design Patterns：Elements of Reusable Object Oriented Software* 的中译本及英文版。

> **工厂方法模式**（别名：虚拟构造）
> 　　定义一个用于创建对象的接口，让子类决定实例化哪一个类。工厂方法模式使一个类的实例化延迟到其子类。
> **Factory Method Pattern**（Another Name：Virtual Constructor）
> 　　Define an interface for creating an object, but let subclasses decide which class to instantiate. Factory Method lets a class defer instantiation to subclasses.

以上内容是 GoF 对工厂方法模式的高度概括，结合 13.2.1 节的工厂方法模式的类图可以准确地理解该模式。

13.1　概述

得到一个类的子类的实例的最常用方法是使用 new 运算符和该子类的构造方法，但是在某些情况下，用户程序不应该或无法使用这个方法来得到一个子类的实例，其原因是系统不允许用户代码和该类的子类形成耦合，或者用户不知道该类有哪些子类可用。例如，"书写工具"有"钢笔""圆珠笔"和"毛笔"等子类，假设这些子类的名字分别是 Pen、BallPen 和 BrushPen。如果用户想得到一支钢笔，系统不想让用户看到钢笔的生产过程之后再给用户一支钢笔（代码相当于 Pen pen=new Pen();），而是让钢笔厂直接给用户一支钢笔（用代码体现就是 Pen pen=penFactory.getWritingTool()），如图 13.1 所示。

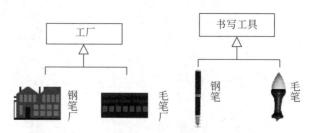

图 13.1　钢笔厂为用户直接提供笔

当系统准备为用户提供某个类的子类的实例，又不想让用户代码和该子类形成耦合（不想让用户代码出现子类的名字），就可以使用工厂方法模式来设计系统。工厂方法模式的关键是在一个接口或抽象类中定义一个抽象方法，该方法返回某个类的子类的实例，该抽象类或接口让其子类或实现该接口的类通过重写这个抽象方法返回某个子类的实例。

13.2 模式的结构与使用

工厂方法模式的结构中包括四种角色。

1. 抽象产品（Product）

抽象产品角色可以是一个抽象类或接口。

2. 具体产品（ConcreteProduct）

具体产品是抽象产品类的子类（如果抽象产品角色是一个抽象类），或实现抽象产品的类（如果抽象产品角色是一个接口）。系统不让用户直接用具体产品的类的构造方法得到具体产品的实例，而是让对应的工厂为用户提供具体产品的实例。

3. 构造者（Creator）

构造者是一个接口或抽象类。构造者负责定义一个被习惯地称为"工厂方法"的抽象方法，该方法的返回类型是抽象产品类型。

4. 具体构造者（ConcreteCreator）

具体构造者是实现构造者的类（如果构造者是接口）或是构造者的子类（如果构造者是抽象类）。具体构造者重写构造者给出的"工厂方法"，以便返回具体产品的实例。具体构造者的实例相当于生产"产品"的工厂，即为用户提供具体产品的实例。

▶ 13.2.1 工厂方法模式的 UML 类图

工厂方法模式的类图如图13.2所示。

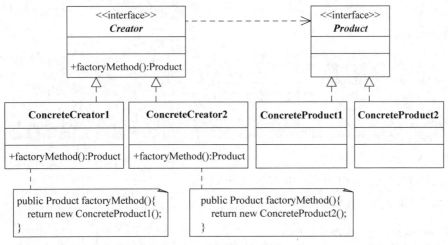

图 13.2　工厂方法模式的类图

13.1节提到的钢笔、毛笔是具体产品的实例，钢笔厂和毛笔厂是具体构造者的实例。

▶ 13.2.2 结构的描述

下面通过一个简单的问题来描述工厂方法模式中所涉及的各个角色。

有很多"书写工具"，例如钢笔、圆珠笔或毛笔等，我们不想让用户看到"书写工具"的生产过程（不想让用户使用 new 得到类的实例），而是直接为用户提供书写工具。请使用工厂方法模式设计几个类，为用户提供几种"书写工具"。

1. 抽象产品

抽象产品角色是 WritingTool 类。代码如下：

WritingTool.java

```
public abstract class WritingTool {  //书写工具
    public abstract void write(String str);
}
```

2. 具体产品

具体产品是 Pen、BallPen 和 BrushPen 三个类，其实例分别模拟钢笔、圆珠笔和毛笔。三个类的代码分别如下：

Pen.java

```
public class Pen extends WritingTool{
    public void write(String str){
        System.out.println("钢笔写出:" + str);
    }
}
```

BallPen.java

```
public class BallPen extends WritingTool{
    public void write(String str){
        System.out.println("圆珠笔写出:" + str);
    }
}
```

BrushPen.java

```
public class BrushPen extends WritingTool{
    public void write(String str){
        System.out.println("毛笔写出:" + str);
    }
}
```

3. 构造者

该角色是 WritingToolFactory 接口。代码如下：

WritingToolFactory.java

```
public interface WritingToolFactory {
    public abstract WritingTool getWritingTool();         //构造者角色中声称的"工厂方法"
}
```

4. 具体构造者

具体构造者角色有三个类，都是 WritingToolFactory 类的子类，分别是 PenFactory 类、BallPenFactory 类和 BrushPenFactory 类。三个类的代码分别如下：

PenFactory.java

```
public class PenFactory implements WritingToolFactory{    //钢笔厂
    public Pen getWritingTool(){
```

```java
        return new Pen();
    }
}
```

BallPenFactory.java

```java
public class BallPenFactory implements WritingToolFactory{    //圆珠笔厂
    public BallPen getWritingTool(){
        return new BallPen();
    }
}
```

BrushPenFactory.java

```java
public class BrushPenFactory implements WritingToolFactory{    //毛笔厂
    public BrushPen getWritingTool(){
        return new BrushPen();
    }
}
```

▶ 13.2.3 模式的使用

前面已经使用工厂方法模式给出了可以使用的类,这些类就是一个小框架,可以使用这个小框架中的类编写应用程序。

下列应用程序(Application.java)使用对应的工厂为用户提供相应的"书写工具"。程序运行效果如图 13.3 所示。

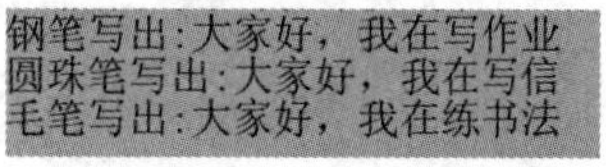

```
钢笔写出:大家好，我在写作业
圆珠笔写出:大家好，我在写信
毛笔写出:大家好，我在练书法
```

图 13.3 程序运行效果

Application.java

```java
public class Application {
    public static void main(String args[]) {
        WritingToolFactory factory = new PenFactory();         //工厂是钢笔厂
        WritingTool myPen = factory.getWritingTool();
        myPen.write("大家好,我在写作业");
        factory = new BallPenFactory();                        //工厂是圆珠笔厂
        myPen = factory.getWritingTool();
        myPen.write("大家好,我在写信");
        factory = new BrushPenFactory();                       //工厂是毛笔厂
        myPen = factory.getWritingTool();
        myPen.write("大家好,我在练书法");
    }
}
```

13.3 工厂方法模式的优点

使用工厂方法可以让用户的代码和某个特定类的子类的代码解耦。工厂方法使用户不必知道它所使用的对象是怎样被创建的,只需要知道该对象有哪些方法即可。

13.4 应用举例——创建药品对象

1. 设计要求

系统目前已经按照有关药品的规定设计了一个抽象类 Medicine，该抽象类特别规定了所创建的药品必须给出药品的成分及其含量。Medicine 目前有两个子类：Paracetamol 和 Amorolfine。Paracetamol 子类负责创建醋氨酚一类的药品；Amorolfine 子类负责创建阿莫罗芬一类的药品。一个为某药店开发的应用程序需要使用 Medicine 类的某个子类的实例为用户提供药品，但是药店的应用程序不能使用 Medicine 的子类的构造方法直接创建对象（不能让客户看见药品的生产过程）。请使用工厂方法模式为用户返回 Medicine 类的子类的实例。

2. 设计实现

1) 抽象产品

抽象产品角色是 Medicine 类。代码如下。

Medicine.java

```
public abstract class Medicine {
    String constitute;
    String name;
    public String getName(){
        return name;
    }
    public String getConstitute(){
        return constitute;
    }
}
```

2) 具体产品

Paracetamol 类（模拟醋氨酚类药品）和 Amorolfine 类（模拟阿莫罗芬类药品）是两个具体产品角色。代码分别如下：

Paracetamol.java

```
public class Paracetamol extends Medicine{
    String part1 = "每粒含乙酰氨基酚";
    String part2 = "每粒含咖啡因";
    String part3 = "每粒含人工牛黄";
    String part4 = "每粒含马来酸氯苯";
    public Paracetamol(String name, int a[]){
        this.name = name;
        part1 = part1 + ":" + a[0] + "毫克\n";
        part2 = part2 + ":" + a[1] + "毫克\n";
        part3 = part3 + ":" + a[2] + "毫克\n";
        part4 = part4 + ":" + a[3] + "毫克\n";
        constitute = part1 + part2 + part3 + part4;
    }
}
```

Amorolfine.java

```java
public class Amorolfine extends Medicine{
    String part1 = "每粒含甲硝唑";
    String part2 = "每粒含人工牛黄";
    public Amorolfine(String name,int a[]){
        this.name = name;
        part1 = part1 + ":" + a[0] + "毫克\n";
        part2 = part2 + ":" + a[1] + "毫克\n ";
        constitute = part1 + part2;
    }
}
```

3）构造者

构造者角色是 MedicineCreator 接口。代码如下：

MedicineCreator.java

```java
public interface MedicineCreator{
    public abstract Medicine getMedicine();     //构造者中的"工厂方法"
}
```

4）具体构造者

具体构造者角色是 ParaMedicineCreator 类和 AmorMedicineCreator 类。代码分别如下：

ParaMedicineCreator.java

```java
public class ParaMedicineCreator implements MedicineCreator{
    public Medicine getMedicine(){
        int a[] = {250,15,1,10};
        Medicine medicine = new Paracetamol("氨加黄敏胶囊",a);
        return medicine;
    }
}
```

AmorMedicineCreator.java

```java
public class AmorMedicineCreator implements MedicineCreator{
    public Medicine getMedicine(){
        int a[] = {200,5};
        Medicine medicine = new Amorolfine("甲硝唑胶囊",a);
        return medicine;
    }
}
```

5）应用程序

前面已经使用工厂方法模式给出了可以使用的类，这些类就是一个小框架，可以使用这个小框架中的类编写应用程序。

下列应用程序（Application.java）使用了上述工厂方法模式中所涉及的抽象产品、构造者以及具体构造者，即使用具体构造者为用户提供药品。程序运行效果如图 13.4 所示。

```
氨加黄敏胶囊的成分：
每粒含乙酰氨基酚:250毫克
每粒含咖啡因:15毫克
每粒含人工牛黄:1毫克
每粒含马来酸氯苯:10毫克

甲硝唑胶囊的成分：
每粒含甲硝唑:200毫克
每粒含人工牛黄:5毫克
```

图 13.4　程序运行效果

Application.java

```java
public class Application{
    public static void main(String args[]){
        MedicineCreator creator = new ParaMedicineCreator();
        Medicine medicine = creator.getMedicine();
        System.out.println(medicine.getName() + "的成分:");
        System.out.println(medicine.getConstitute());
        creator = new AmorMedicineCreator();
        medicine = creator.getMedicine();
        System.out.println(medicine.getName() + "的成分:");
        System.out.println(medicine.getConstitute());
    }
}
```

第 14 章 抽象工厂模式

以下文本框中的内容引自 GoF 所著 *Design Patterns*：*Elements of Reusable Object Oriented Software* 的中译本及英文版。

> **抽象工厂模式（别名：配套）**
> 提供一个创建一系列相关或相互依赖对象的接口,而无须指定它们具体的类。
> **Abstract Factory Pattern（Another Name：Kit）**
> Provide an interface for creating families of related or dependent objects without specifying their concrete classes.

以上内容是 GoF 对抽象工厂模式的高度概括,结合 14.2.1 节的抽象工厂模式的类图可以准确地理解该模式。

14.1 概述

设计某些系统时可能需要为用户提供一系列相关的对象,但系统不希望用户直接使用 new 运算符实例化这些对象,而是由系统来控制这些对象的创建,否则用户不仅要清楚地知道使用哪些类来创建这些对象,而且还必须清楚这些对象之间是如何相关的,使得用户的代码和这些类形成紧耦合,缺乏弹性,不利于维护。例如,军队(系统)要为士兵(用户)提供机枪、手枪以及相应的子弹,但军队(系统)不希望由士兵(用户)来生产机枪、手枪及相应的子弹(不使用 new 运算符),而是有专门的工厂负责配套生产,即有一个专门负责生产机枪、机枪子弹的工厂和一个专门负责生产手枪、手枪子弹的工厂,如图 14.1 所示。

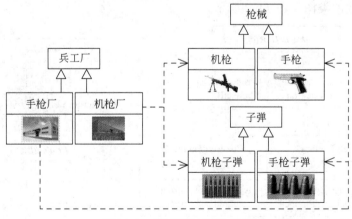

图 14.1 兵工厂与枪械、子弹

当系统准备为用户提供一系列相关的对象,又不想让用户代码和创建这些对象的类形成耦合时,就可以使用抽象工厂模式来设计系统。抽象工厂模式的关键是在一个抽象类或接口中定义若干个抽象方法,这些抽象方法分别返回某个类的实例,该抽象类或接口让其子类或实现该接口的类重写这些抽象方法,为用户提供一系列相关的对象。

14.2 模式的结构与使用

抽象工厂模式的结构中包括四种角色。

1. 抽象产品(Product)

抽象产品的角色里可以包含多个抽象类或接口。

2. 具体产品(ConcreteProduct)

具体产品是抽象产品类的子类(如果抽象产品角色是一个抽象类),或实现抽象产品的类(如果抽象产品角色是一个接口)。系统不让用户直接用具体产品的类的构造方法得到具体产品的实例,而是让对应的具体工厂为用户提供具体产品的实例。

3. 抽象工厂(AbstractFactory)

抽象工厂角色是一个接口或抽象类,负责定义返回一系列对象的抽象方法,这些方法的返回类型都是抽象产品类型。

4. 具体工厂(ConcreteFactory)

如果抽象工厂是抽象类,具体工厂是抽象工厂的子类;如果抽象工厂是接口,具体工厂是实现抽象工厂的类。具体工厂重写抽象工厂中的全部抽象方法,让每个重写的抽象方法返回一个具体产品的实例。重写的全部抽象方法所返回的各个实例之间有依赖关系。

> **注意**:抽象工厂模式和工厂方法模式很类似,但是抽象工厂模式中的具体工厂角色要负责返回一系列相关的对象(一系列相关的具体产品的实例),而工厂方法模式中的具体构造者角色仅仅负责返回某个子类的对象(一个具体产品的实例)。

▶ 14.2.1 抽象工厂模式的 UML 类图

抽象工厂模式的类图如图 14.2 所示。

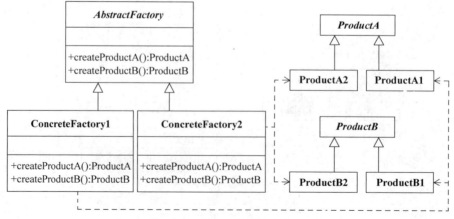

图 14.2 抽象工厂模式的类图

14.1节提到的"枪械"和"子弹"是模式中的抽象产品角色,"机枪""机枪子弹""手枪""手枪子弹"是具体产品角色;"兵工厂"是抽象工厂角色,"机枪厂"和"手枪厂"是具体工厂角色。按照 UML 图,机枪厂负责生产机枪和相关的机枪子弹,手枪厂负责生产手枪和相关的手枪子弹。

▶ 14.2.2 结构的描述

下面通过一个简单的问题来描述抽象工厂模式中所涉及的各个角色。

要生产机枪及配套的机枪子弹、手枪及配套的手枪子弹,请使用抽象工厂模式建立一个系统,该系统可以为用户提供机枪和机枪子弹、手枪和手枪子弹。

1. 抽象产品

抽象产品角色有两个类,分别是 Gun 类(枪的抽象描述)和 Bullet 类(子弹的抽象描述),给出了具体产品需要实现的抽象方法。代码分别如下:

Gun. java

```java
import java.util.Stack;
public abstract class Gun{                                    //枪
    public Stack<Bullet> clip;                                //弹夹
    public abstract String getModel();                        //型号
    public abstract void loadingBullet(Bullet ...bullet);     //装弹
    public abstract void fireShotting();                      //射击
}
```

Bullet. java

```java
public abstract class Bullet {
    public abstract String getBulletType();                   //子弹的型号
}
```

2. 具体产品

具体产品是抽象产品的子类,共有四个类,分别是用来模拟机枪的 MachineGun 类(抽象类 Gun 的一个子类)、模拟手枪的 Pistol 类(抽象类 Gun 的一个子类)、模拟机枪子弹的 MachineGunBullet 类(抽象类 Bullet 的一个子类)、模拟手枪子弹的 PistolBullet 类(抽象类 Bullet 的一个子类)。需要注意的是,MachineGun 类的实例只能使用 MachineGunBullet 类的实例(机枪只能使用机枪子弹),Pistol 类的实例只能使用 PistolBullet 类的实例(手枪只能使用手枪子弹),只有这样才符合抽象工厂模式的结构。

在阅读代码时要注意一个 Java 的知识点(见第 2 章),如果 B 是 A 的类,对象 a 是子类 B 的对象的上转型对象,即"A a=new B();",那么表达式"a instanceof B"的值是 true。

四个类的代码分别如下:

MachineGun. java

```java
import java.util.Stack;
public class MachineGun extends Gun {                         //机枪
    public MachineGun(){
        clip = new Stack<Bullet>();
    }
    public String getModel() {                                //型号
        return "机枪";
```

```java
    }
    public void loadingBullet(Bullet ...bullet){
        for(Bullet b:bullet){
            if(b instanceof MachineGunBullet) {        //如果是机枪子弹
                if(clip.size()<=100)                    //最大装弹量是100发
                    clip.push(b);
                else {
                    System.out.print("枪是" + getModel() + ",");
                    System.out.println("子弹是" + b.getBulletType() + "!无法装弹。");
                }
            }
        }
    }
    public void fireShotting(){
        if(!clip.empty()) {
            System.out.println("有子弹!");
            System.out.print(getModel() + ":哒...哒...射出子弹:");
            for(int i = 0;i < 5;i++){
                if(!clip.empty())
                    System.out.print(clip.pop().getBulletType());
            }
            System.out.println();
        }
        else {
            System.out.println(getModel() + "没有子弹,需要装弹。");
        }
    }
}
```

Pistol.java

```java
import java.util.Stack;
public class Pistol extends Gun {                      //手枪
    public Pistol(){
        clip = new Stack<Bullet>();
    }
    public String getModel() {                          //型号
        return "手枪";
    }
    public void loadingBullet(Bullet ...bullet){
        for(Bullet b:bullet){
            if(b instanceof PistolBullet) {            //如果是手枪子弹
                if(clip.size()<=20)                     //最大装弹量是20发
                    clip.push(b);
                else {
                    System.out.print("枪是" + getModel() + ",");
                    System.out.println("子弹是" + b.getBulletType() + "!无法装弹。");
                }
            }
        }
    }
```

```java
    public void fireShotting(){
        if(!clip.empty()) {
            System.out.println("有子弹!");
            System.out.print(getModel() + ":射出 1 发子弹:");
            System.out.println(clip.pop().getBulletType());
        }
        else {
            System.out.println(getModel() + "没有子弹,需要装弹。");
        }
    }
}
```

MachineGunBullet. java

```java
public class MachineGunBullet extends Bullet{           //机枪子弹
    public String getBulletType(){                      //子弹的型号
        return "●";
    }
}
```

PistolBullet. java

```java
public class PistolBullet extends Bullet{               //手枪子弹
    public String getBulletType(){                      //子弹的型号
        return "◎";
    }
}
```

3. 抽象工厂

抽象工厂角色是 MilitaryFactory 类,模拟军工厂,MilitaryFactory 类中的两个抽象方法的返回类型分别是 Gun 和 Bullet。代码如下:

MilitaryFactory. java

```java
public abstract class MilitaryFactory {                 //军工厂
    public abstract Gun manufacturingGun();             //制造枪
    public abstract Bullet manufacturingBullet();       //制造子弹
}
```

4. 具体工厂

该角色有两个类,分别是 MachineGunFactory 类(模拟机枪厂)和 PistolFactory 类(模拟手枪厂)。MachineGunFactory 类的实例负责返回 MachineGun 类和 MachineGunBullet 类的实例,用来模拟制造机枪和机枪子弹;PistolFactory 类的实例负责返回 PistolGun 类和 PistolBullet 类的实例,用来模拟制造手枪和手枪子弹。代码分别如下:

MachineGunFactory. java

```java
public class MachineGunFactory extends MilitaryFactory {   //机枪厂
    public Gun manufacturingGun(){                         //制造机枪
        return new MachineGun();
    }
    public Bullet manufacturingBullet(){                   //制造机枪子弹
```

```java
        return new MachineGunBullet();
    }
}
```

PistolFactory.java

```java
public class PistolFactory extends MilitaryFactory {          //手枪厂
    public Gun manufacturingGun(){                            //制造手枪
        return new Pistol();
    }
    public Bullet manufacturingBullet(){                      //制造手枪子弹
        return new PistolBullet();
    }
}
```

▶ 14.2.3 模式的使用

前面已经使用抽象工厂模式给出了可以使用的类,这些类就是一个小框架,可以使用这个小框架中的类编写应用程序。

下列应用程序(Application.java)使用对应的工厂为用户提供相应的机枪和机枪子弹、手枪和手枪子弹。程序运行效果如图 14.3 所示。

```
目前的枪是机枪。
机枪没有子弹,需要装弹。
有子弹!
机枪:哒...哒...射出子弹:●●●●●
有子弹!
机枪:哒...哒...射出子弹:●●●
机枪没有子弹,需要装弹。
枪是机枪,子弹是◎!无法装弹。
目前的枪是手枪。
手枪没有子弹,需要装弹。
有子弹!
手枪:射出1发子弹:◎
手枪没有子弹,需要装弹。
枪是手枪,子弹是●!无法装弹。
```

图 14.3 程序运行效果

Application.java

```java
public class Application {
    public static void main(String args[]) {
        MilitaryFactory factory = new MachineGunFactory();          //机枪厂
        Gun gun = factory.manufacturingGun();                       //gun 是机枪
        System.out.println("目前的枪是" + gun.getModel() + "。");
        gun.fireShotting();
        for(int i = 1;i <= 8;i++){
            gun.loadingBullet(factory.manufacturingBullet());
        }
        gun.fireShotting();
        gun.fireShotting();
        gun.fireShotting();
        factory = new PistolFactory();                              //手枪厂
        gun.loadingBullet(factory.manufacturingBullet());           //机枪无法装手枪子弹
        gun = factory.manufacturingGun();                           //gun 是手枪(换枪了)
```

```
            System.out.println("目前的枪是" + gun.getModel() + "。");
            gun.fireShotting();
            gun.loadingBullet(factory.manufacturingBullet());
            gun.fireShotting();
            gun.fireShotting();
            factory = new MachineGunFactory();                    //机枪厂
            gun.loadingBullet(factory.manufacturingBullet());     //手枪无法装机枪子弹
    }
}
```

14.3 抽象工厂模式的优点

抽象工厂模式具有以下优点：

（1）抽象工厂模式可以为用户创建一系列相关的对象，让用户和创建这些对象的类脱耦。

（2）使用抽象工厂模式可以方便地为用户配置一系列对象。用户使用不同的具体工厂就能得到一组相关的对象，同时也能避免用户混用不同系列中的对象。

（3）在抽象工厂模式中，可以随时增加具体工厂为用户提供一组相关的对象。

14.4 应用举例——存款凭证

扫一扫

视频讲解

1. 设计要求

用户在银行存款后，将得到银行给予的存款凭证，该存款凭证就是加盖了业务公章的存款明细。不同银行的业务公章不仅名称互不相同，而且形状也互不相同，例如交通银行的业务公章是正方形，中国银行的业务公章是圆形，中国建设银行的业务公章是等边三角形。请使用抽象工厂模式为用户提供存款凭证。

2. 设计实现

1）抽象产品

该角色有两个接口，分别是 Deposit 和 Stamp，分别用来刻画"存款明细"和"业务公章"。二者的代码分别如下：

Deposit.java

```java
public interface Deposit {                                //存单接口
    public abstract String getBankName();                 //返回银行的名称
    public abstract String getClientName();               //返回存单的客户名称
    public abstract String getSlipNumber();               //返回存单的存单号
    public abstract int getAmountOfMoney();               //返回存单的金额
    public abstract void setStamp(Stamp stamp);           //设置存单公章
    public abstract Stamp getStamp();                     //返回存单公章
}
```

Stamp.java

```java
import java.awt.Image;
public interface Stamp {                                  //业务公章接口
    public abstract Image getImage();                     //返回一幅图像
}
```

2）具体产品

在阅读代码时要注意一个 Java 的知识点（见第 2 章），如果 B 是 A 的类，对象 a 是子类 B 的对象的上转型对象，即"A a=new B();"，那么表达式"a instanceof B"的值是 true。

该角色有四个类，分别是 BankOfChinaDeposit（实现 Deposit 接口，模拟中国银行的存单）、ICBCDeposit（实现 Deposit 接口，模拟中国工商银行的存单）、BankOfChinaStamp（实现 Stamp 接口，模拟中国银行的公章）和 ICBCStamp（实现 Stamp 接口，模拟中国工商银行的公章）。需要注意的是，BankOfChinaDeposit 类的实例只能使用 BankOfChinaStamp 类的实例（中国银行的存单只能使用中国银行的业务公章），ICBCDeposit 类的实例只能使用 ICBCStamp 类的实例（中国工商银行的存单只能使用中国工商银行的业务公章），只有这样才符合抽象工厂模式的结构。四个类的代码分别如下：

BankOfChinaDeposit.java

```java
public class BankOfChinaDeposit implements Deposit {          //中国银行的存单
    String clientName;                                         //客户名称
    String slipNumber;                                         //存单号
    int money;                                                 //存单金额
    Stamp stamp;                                               //公章
    public BankOfChinaDeposit(String slipNumber,String clientName,int money){
        this.slipNumber = slipNumber;
        this.clientName = clientName;
        this.money = money;
    }
    public void setStamp(Stamp stamp) {
        if(stamp instanceof BankOfChinaStamp) {                //必须是本银行的业务公章
            this.stamp = stamp;
        }
        else {
            System.out.println("请使用正确的" + getBankName() + "公章");
        }
    }
    public String getBankName(){
        return "中国银行";
    }
    public String getClientName(){
        return clientName;
    }
    public String getSlipNumber(){
        return slipNumber;
    }
    public int getAmountOfMoney(){
        return money;
    }
    public Stamp getStamp(){
        return stamp;
    }
}
```

第14章 抽象工厂模式

ICBCDeposit.java

```java
public class ICBCDeposit implements Deposit {          //中国工商银行的存单
    String clientName;                                  //客户名称
    String slipNumber;                                  //存单号
    int money;                                          //存单金额
    Stamp stamp;                                        //公章
    public ICBCDeposit(String slipNumber,String clientName,int money){
        this.slipNumber = slipNumber;
        this.clientName = clientName;
        this.money = money;
    }
    public void setStamp(Stamp stamp) {
        if(stamp instanceof ICBCStamp) {                //必须是本银行的业务公章
            this.stamp = stamp;
        }
        else {
            System.out.println("请使用正确的" + getBankName() + "公章");
        }
    }
    public String getBankName(){
        return "中国工商银行";
    }
    public String getClientName(){
        return clientName;
    }
    public String getSlipNumber(){
        return slipNumber;
    }
    public int getAmountOfMoney(){
        return money;
    }
    public Stamp getStamp(){
        return stamp;
    }
}
```

BankOfChinaStamp.java

```java
import java.awt.Image;
import java.awt.Toolkit;
public class BankOfChinaStamp implements Stamp{        //中国银行的业务公章
    public Image getImage(){                            //返回一幅图像
        Toolkit tool = Toolkit.getDefaultToolkit();     //tool用于获取图像
        Image img = tool.getImage("bankOfChinaStamp.jpg");
        return img;
    }
}
```

ICBCStamp.java

```java
import java.awt.Image;
import java.awt.Toolkit;
```

```java
public class ICBCStamp implements Stamp{           //中国工商银行的业务公章
    public Image getImage(){                        //返回一幅图像
        Toolkit tool = Toolkit.getDefaultToolkit(); //tool用于获取图像
        Image img = tool.getImage("ICBCStamp.jpg");
        return img;
    }
}
```

3）抽象工厂

抽象工厂角色是 Bank 类，模拟银行，Bank 类中的两个抽象方法的返回类型分别是 Deposit 和 Stamp。代码如下：

Bank.java

```java
public abstract class Bank {                        //银行
    public abstract Deposit getDeposit(String clientNumber,String slipNumber,int money);
                                                    //返回存单
    public abstract Stamp getStamp();               //返回公章
}
```

4）具体工厂

该角色有两个类，分别是 BankOfChina 类（模拟中国银行）和 ICBC 类（模拟中国工商银行）。BankOfChina 类的实例负责返回 BankOfChinaDeposit 类和 BankOfChinaStamp 类的实例；ICBC 类的实例负责返回 ICBCDeposit 类和 ICBCStamp 类的实例。两个类的代码分别如下：

BankOfChina.java

```java
public class BankOfChina extends Bank {             //中国银行
    public Deposit getDeposit(String clientNumber,String slipNumber,int money){
                                                    //返回存单
        return new BankOfChinaDeposit(clientNumber,slipNumber,money);
    }
    public Stamp getStamp(){                        //返回公章
        return new BankOfChinaStamp();
    }
}
```

ICBC.java

```java
public class ICBC extends Bank {                    //中国工商银行
    public Deposit getDeposit(String clientNumber,String slipNumber,int money){
                                                    //返回存单
        return new ICBCDeposit(clientNumber,slipNumber,money);
    }
    public Stamp getStamp(){                        //返回公章
        return new ICBCStamp();
    }
}
```

5）应用程序

前面已经使用抽象工厂模式给出了可以使用的类，这些类就是一个小框架，可以使用这个小框架中的类编写应用程序。

第14章 抽象工厂模式

实际生活中,客户在银行存款就可得到存单,用户不能越过银行直接和存单打交道(抽象工厂模式的关键一点就是:用户不能直接用具体产品角色中的类实例化对象),即应用程序只和 Bank 的某些子类打交道,即可得到相应的实现 Deposit 接口类的实例。下列应用程序(Application.java)给出了中国银行和中国工商银行的存单。程序运行效果如图 14.4 所示。

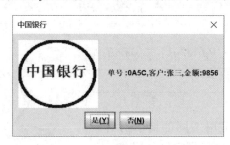 

图 14.4　程序运行效果

Application.java

```java
import javax.swing.JOptionPane;
import javax.swing.ImageIcon;
public class Application {
    public static void main(String args[]) {
        Bank bank = new BankOfChina();                              //银行是中国银行
        Deposit deposit = bank.getDeposit("0A5C","张三",9856);      //存单是中国银行存单
        Stamp stamp = bank.getStamp();                              //公章是中国银行公章
        String bankName = deposit.getBankName();                    //存单上的银行名称
        String clientName = deposit.getClientName();                //存单上的客户名称
        String slipNumber = deposit.getSlipNumber();                //存单上的号码
        int money = deposit.getAmountOfMoney();                     //存单上的金额
        ImageIcon icon = new ImageIcon(stamp.getImage());
        JOptionPane.showConfirmDialog(null,
                        "单号 :" + slipNumber + ",客户:" + clientName + ",金额:" + money,
                        bankName,
                        JOptionPane.YES_NO_OPTION,
                        JOptionPane.INFORMATION_MESSAGE,
                        icon);
        bank = new ICBC();                                          //银行是中国工商银行
        deposit = bank.getDeposit("518D","李四",19759);             //存单是中国工商银行存单
        stamp = bank.getStamp();                                    //公章是中国工商银行公章
        bankName = deposit.getBankName();                           //存单上的银行名称
        clientName = deposit.getClientName();                       //存单上的客户名称
        slipNumber = deposit.getSlipNumber();                       //存单上的号码
        money = deposit.getAmountOfMoney();                         //存单上的金额
        icon = new ImageIcon(stamp.getImage());
        JOptionPane.showConfirmDialog(null,
                        "单号 :" + slipNumber + ",客户:" + clientName + ",金额:" + money,
                        bankName,
                        JOptionPane.YES_NO_OPTION,
                        JOptionPane.INFORMATION_MESSAGE,
                        icon);
        System.exit(0);
    }
}
```

第 15 章　命令模式

以下文本框中的内容引自 GoF 所著 *Design Patterns*：*Elements of Reusable Object Oriented Software* 的中译本及英文版。

命令模式（别名：动作，事务）
　　将一个请求封装为一个对象，从而使用户可用不同的请求对客户进行参数化；对请求排队或记录请求日志，以及支持可撤销的操作。
Command Pattern（Another Name：Action，Transaction）
　　Encapsulate a request as an object, thereby letting you parameterize clients with different requests, queue or log requests, and support undoable operations.

以上内容是 GoF 对命令模式的高度概括，结合 15.2.1 节的命令模式的类图可以准确地理解该模式。

15.1　概述

在许多设计中，经常涉及一个对象请求另一个对象调用其方法达到某种目的。如果请求者不希望或无法直接和被请求者打交道，即不希望或无法含有被请求者的引用，那么就可以使用命令模式。例如，在军队作战中，指挥官请求三连偷袭敌人，但是指挥官不希望或无法直接与三连取得联系，那么可以将"三连偷袭敌人"这个请求形成一个作战命令，该作战命令的核心就是"三连偷袭敌人"。只要能让该作战命令被执行（即使指挥官已经不存在），就会实现三连偷袭敌人的目的，如图 15.1 所示。

图 15.1　军官请求三连偷袭敌人

命令模式是关于怎样处理一个对象请求另一个对象调用其方法完成某项任务的一种成熟的模式，这里称提出请求的对象为请求者，被请求的对象为接收者。在命令模式中，请求者不和接收者直接打交道，而是把这种"请求"封装到一个称作"命令"的对象中。这样一来，当请求者让"命令"执行方法时，"命令"就会让接收者执行某个方法。

15.2 模式的结构与使用

命令模式的结构中包括 4 种角色。

1. 接收者(Receiver)

接收者角色包含若干个类,每个类的实例都是一个接收者。

2. 命令(Command)

命令角色是一个接口,规定了若干个抽象方法,例如 execute()、undo()等。命令接口中的方法的主要职责是让接收者执行某些操作,即执行某些方法。命令接口还可以提供用来撤销的方法(不是必需的),例如 undo()方法,即 undo()方法的执行能撤销 execute()方法的执行效果。如果 execute()方法的执行效果不可撤销(例如退出程序等),那么"具体命令"角色中的类在重写 undo()方法时,可以用"空实现"的方式。

3. 具体命令(ConcreteCommand)

具体命令角色是实现命令接口的类,其实例称为命令。

4. 请求者(Invoker)

请求者角色是一个类,该类含有 Command 接口变量。该类的实例称为请求者,请求者和命令打交道(命令和接收者打交道,即含有接收者的引用)。

15.2.1 命令模式的 UML 类图

命令模式的类图如图 15.2 所示。

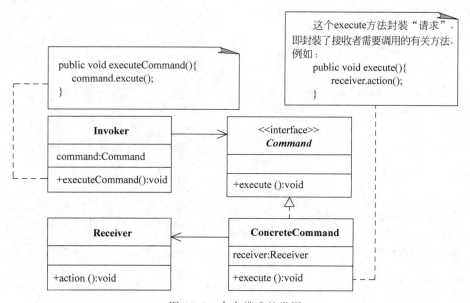

图 15.2 命令模式的类图

15.2.2 结构的描述

下面通过一个简单的问题来描述命令模式中所涉及的各个角色。

一个指挥官命令三连偷袭敌人、一团开始渡河、工兵连在无名路口布雷四颗,但 5 分钟后,又撤销了给工兵连的命令。

1. 接收者

接收者角色有 ArmyBeAttack、ArmyBeCrossingRiver 和 ArmyBeMine 三个类。ArmyBeAttack 类的实例用来模拟负责偷袭的部队；ArmyBeCrossingRiver 类的实例用来模拟负责渡河的部队；ArmyBeMine 类的实例用来模拟负责布雷的部队。三个类的代码分别如下：

ArmyBeAttack.java

```java
public class ArmyBeAttack {
    String armyName;
    ArmyBeAttack(){
        armyName = "";
    }
    ArmyBeAttack(String name){
        armyName = name;
    }
    public void sneakAttack(){
        System.out.println(armyName + "开始偷袭敌人...");
    }
}
```

ArmyBeCrossingRiver.java

```java
public class ArmyBeCrossingRiver{
    String armyName;
    ArmyBeCrossingRiver(){
        armyName = "";
    }
    ArmyBeCrossingRiver(String name){
        armyName = name;
    }
    public void crossingRiver(){
        System.out.println(armyName + "开始渡河...");
    }
}
```

ArmyBeMine.java

```java
public class ArmyBeMine{
    String armyName;
    ArmyBeMine(){
        armyName = "";
    }
    ArmyBeMine(String name){
        armyName = name;
    }
    public void layMine(){                    //模拟布雷
        System.out.print("●●●●●");
    }
    public void clearMine() {                 //模拟清除雷
        System.out.print("\r");               //退行(回到行首,但不进行换行)
        System.out.println("○○○○○");        //输出 5 个○,模拟排雷
```

```
        System.out.print("排雷结束");
    }
}
```

2. 命令

这里的命令是一个名字为 Command 的接口,该接口给出了用来封装"请求"的方法。代码如下:

Command.java

```
public interface Command {
    public abstract void execute();
    public abstract void undo();
}
```

3. 具体命令

具体命令角色是实现命令接口的类,本问题中有三个实现 Command 接口的类,分别是 AttackCommand、CrossingRiverCommand 和 MineCommand。其中,AttackCommand 类模拟"偷袭命令";CrossingRiverCommand 类模拟"渡河命令";MineCommand 类模拟"布雷命令"。三个类的代码分别如下:

AttackCommand.java

```
public class AttackCommand implements Command {
    ArmyBeAttack army;
    AttackCommand(ArmyBeAttack army){
        this.army = army;
    }
    public void execute(){            //封装着指挥官的请求
        army.sneakAttack();           //部队偷袭
    }
    public void undo(){
        //偷袭已经开始,无法撤销(重写用空实现)
    }
}
```

CrossingRiverCommand.java

```
public class CrossingRiverCommand implements Command {
    ArmyBeCrossingRiver army;
    CrossingRiverCommand(ArmyBeCrossingRiver army){
        this.army = army;
    }
    public void execute(){            //封装着指挥官的请求
        army.crossingRiver();         //部队渡河
    }
    public void undo(){
        //渡河已经开始,无法撤销(重写用空实现)
    }
}
```

MineCommand.java

```java
public class MineCommand implements Command {
    ArmyBeMine army;
    MineCommand(ArmyBeMine army){
        this.army = army;
    }
    public void execute(){                //封装着指挥官的请求
        army.layMine();                   //部队布雷
    }
    public void undo(){
        army.clearMine();                 //部队清除雷
    }
}
```

4．请求者

请求者角色是 ArmySuperior 类，其实例模拟一个指挥官。代码如下：

ArmySuperior.java

```java
public class ArmySuperior{
    Command command;                      //用来存放命令的引用
    public void setCommand(Command command){
        this.command = command;
    }
    public void startExecuteCommand(){
        command.execute();                //让命令执行 execute()方法
    }
    public void cancelCommand(){
        command.undo();                   //让命令执行 undo()方法
    }
}
```

15.2.3 模式的使用

前面已经使用命令模式给出了可以使用的类，这些类就是一个小框架，可以使用这个小框架中的类编写应用程序。

下列应用程序（Application.java）演示了一个指挥官下达命令的过程。程序运行效果如图15.3所示。

图 15.3 程序运行效果

Application.java

```java
public class Application{
    public static void main(String args []){
        var armyCompany = new ArmyBeAttack("三连");              //创建接收者
        var armyGroup = new ArmyBeCrossingRiver("一团");
        var armyEngineer = new ArmyBeMine("工兵连");
        Command commandOne = new AttackCommand(armyCompany);     //具体命令并指定接收者
        Command commandTwo = new CrossingRiverCommand(armyGroup);
        Command commandThree = new MineCommand(armyEngineer);
        ArmySuperior chief = new ArmySuperior();                 //创建指挥官(请求者)
```

```
            chief.setCommand(commandOne);
            chief.startExecuteCommand();
            chief.setCommand(commandTwo);
            chief.startExecuteCommand();
            chief.setCommand(commandThree);
            chief.startExecuteCommand();
            try{
                Thread.sleep(5000);         //模拟5秒后,取消命令
            }
            catch(Exception exp){}
            chief.cancelCommand();
    }
}
```

15.3　命令模式的优点

命令模式具有以下优点:

(1) 在命令模式中,请求者(Invoker)不直接与接收者(Receiver)交互,即请求者(Invoker)不包含接收者(Receiver)的引用,因此彻底消除了彼此之间的耦合。

(2) 命令模式满足"开-闭"原则。如果增加新的具体命令和该命令的接收者,不必修改调用者的代码,调用者就可以使用新的命令对象;反之,如果增加新的调用者,不必修改现有的具体命令和接收者,新增加的调用者就可以使用已有的具体命令。

(3) 由于请求者的请求被封装到了具体命令中,那么就可以将具体命令保存到持久化的媒介中,在需要的时候,重新执行这个具体命令。因此,使用命令模式可以记录日志。

(4) 使用命令模式可以对请求者的"请求"进行排队。每个请求都各自对应一个具体命令,因此可以按一定顺序执行这些具体命令。

15.4　应用举例——开灯与关灯

扫一扫

视频讲解

1. 设计要求

借助javax.swing包提供的组件,使用命令模式模拟开灯与关灯。

2. 设计实现

1) 接收者

接收者角色是Light类,其实例模拟照明灯。代码如下:

Light.java

```
import javax.swing.*;
public class Light extends JPanel{
    Icon imageIcon;
    JLabel label;
    public Light(){
        label = new JLabel();
        add(label);
    }
```

```java
    public void on(){
        label.setIcon(new ImageIcon("lightOpen.jpg"));
    }
    public void off(){
        label.setIcon(new ImageIcon("lightClose.jpg"));
    }
}
```

2) 命令接口

命令接口是 Command 接口。对于某些问题，一些具体命令有着固定的顺序，即这些命令是按顺序被依次执行的，或这些命令是被循环执行的。例如，对于本问题，开灯命令和关灯命令就构成循环命令，因此，命令接口的 execute() 方法的返回类型应该是 Command 类型，以方便具体命令之间的切换。Command 接口代码如下：

Command.java

```java
public interface Command {
    public abstract Command execute();
    public abstract void setNextCommand(Command nextCommand);
}
```

3) 具体命令

具体命令中有两个类，分别是 OnLightCommand 和 OffLightCommand，其实例模拟"开灯命令"和"关灯命令"。代码分别如下：

OnLightCommand.java

```java
public class OnLightCommand implements Command{
    Light light;
    Command nextCommand;                    //当前命令的下一个命令
    OnLightCommand(Light light){
        this.light = light;
    }
    public void setNextCommand(Command nextCommand) {
        this.nextCommand = nextCommand;
    }
    public Command execute(){
        light.on();
        return nextCommand;
    }
}
```

OffLightCommand.java

```java
public class OffLightCommand implements Command{
    Light light;
    Command nextCommand;                    //当前命令的下一个命令
    OffLightCommand(Light light){
        this.light = light;
    }
    public void setNextCommand(Command nextCommand) {
```

```
        this.nextCommand = nextCommand;
    }
    public Command execute(){
        light.off();
        return nextCommand;
    }
}
```

4）请求者

请求者是 Invoke 类，该类的实例含有一个按钮组件（模拟开关），单击按钮来执行相应的命令。每次单击按钮后都导致请求者所组合的命令发生变化，即执行完"开灯命令"，下一个要执行的命令自然是"关灯命令"。代码如下：

Invoke.java

```
import javax.swing.JButton;
import javax.swing.JPanel;
public class Invoke extends JPanel {
    boolean init = false;
    JButton buttonOnAndOff;                    //需要的按钮
    Command command;
    Invoke(){
        buttonOnAndOff = new JButton("开/关");
        add(buttonOnAndOff);
        buttonOnAndOff.addActionListener((e) ->{
            command = command.execute();
        });
    }
    public void setCommand(Command command){
        if(init == false)
            this.command = command;
        init = true;
    }
}
```

5）应用程序

前面已经使用命令模式给出了可以使用的类，这些类就是一个小框架，可以使用这个小框架中的类编写应用程序。

下列应用程序（Application.java）使用命令模式中的接收者（模拟灯）和请求者（模拟开关）模拟开灯和关灯。程序运行效果如图 15.4 所示。

图 15.4　程序运行效果

Application.java

```java
import javax.swing.*;
import java.awt.*;
public class Application {
    public static void main(String args[]) {
        Light light = new Light();                              // 接收者
        Command commandOn = new OnLightCommand(light);          //开灯命令
        Command commandOff = new OffLightCommand(light);        //关灯命令
        commandOn.setNextCommand(commandOff);
        commandOff.setNextCommand(commandOn);
        Invoke invoke = new Invoke();                           //请求者
        invoke.setCommand(commandOn);
        JFrame win = new JFrame();
        win.setBounds(12,12,400,300);
        win.add(light,BorderLayout.CENTER);
        win.add(invoke,BorderLayout.SOUTH);
        win.setDefaultCloseOperation(JFrame.EXIT_ON_CLOSE);
        win.setVisible(true);
    }
}
```

第 16 章　桥接模式

以下文本框中的内容引自 GoF 所著 *Design Patterns：Elements of Reusable Object Oriented Software* 的中译本及英文版。

> **桥接模式**（别名：柄体模式）
> 　将抽象部分与它的实现部分分离，使它们都可以独立地变化。
> **Bridge Pattern**(Another Name：Handle Body)
> 　Decouple an abstraction from its implementation so that the two can vary independently.

以上内容是 GoF 对桥接模式的高度概括，结合 16.2.1 节的桥接模式的类图可以准确地理解该模式。

16.1　概述

抽象类或接口中可以定义若干个抽象方法，习惯上将抽象方法称作操作。抽象类或接口使程序的设计者忽略操作的细节，即不必考虑这些操作是如何实现的，当用户程序面向抽象类或接口时，就不会依赖具体的实现，使系统有很好的扩展性（见第 4 章）。但是，抽象类中的抽象方法总归是需要子类去实现的，在大多数情况下，抽象类的子类完全可以胜任这样的工作，只是在某些情况下，子类可能会遇到一些难以处理的问题，例如，某个编辑要策划出版有关程序设计方面的图书，那么编辑就要按照以下步骤（这里的步骤相当于抽象类给出的抽象方法）完成有关策划工作。

① setISBN()：确定书号

② setBookContent()：确定书中的内容

步骤①、②是图书策划中的高层次步骤，步骤①由编辑负责完成，但步骤②的工作无法由编辑自己独立完成，因为该步骤里包含细节，即图书的内容，完成图书的内容属于编辑策划中的基本操作（细节），应该把细节交给作者来完成，如图 16.1 所示。因此，在步骤②中，编辑应该让作者来完成图书内容的编写，即让作者来完成如下的基本操作：

② setBookContent()→authorWritingBook()

图 16.1　编辑与作者

桥接模式是关于怎样将抽象部分与它的实现部分分离，使它们都可以独立地变化的成熟模式。即抽象类的子类在重写抽象方法时，可以将方法的实现部分再交给另外一个接口，即交给另外一个实现接口的类的实例去完成。

16.2 模式的结构与使用

桥接模式包括四种角色。

1. 抽象（Abstraction）

抽象角色是一个抽象类，该抽象类含有实现者的引用，即含有实现者（Implementor）声明的变量。在抽象角色中，称抽象角色为"抽象层次"，抽象类中的抽象方法是桥接模式中的"高层次操作"。

2. 实现者（Implementor）

实现者角色是一个接口（抽象类）。接口（抽象类）中的方法不一定与抽象角色中的抽象类中的方法一致。称实现者角色为"实现层次"，接口（抽象类）中的抽象方法是桥接模式中的"基本操作"。

3. 细化抽象（Refined Abstraction）

细化抽象是抽象角色中的抽象类的一个子类，该子类在重写（覆盖）抽象类中的抽象方法时，会委托实现者调用相应的方法（将抽象部分与它的实现部分分离）。

4. 具体实现者（Concrete Implementor）

具体实现者是实现了实现者的类（如果实现者是一个接口），或是实现者的一个子类（如果实现者是一个抽象类）。

▶ 16.2.1 桥接模式的 UML 类图

桥接模式的类图如图 16.2 所示。

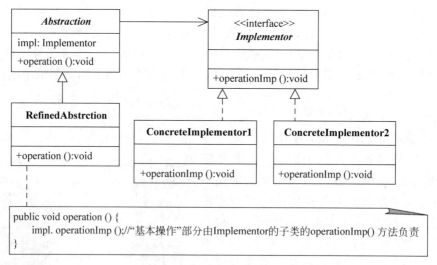

图 16.2 桥接模式的类图

在桥接模式中，将抽象角色和细化抽象角色称为"抽象层次"，将实现者和具体实现者称为"实现层次"。抽象层次负责抽象数据处理，实现层次负责数据的细化处理；抽象层次的操作被称为高层次操作，实现层次的操作被称为基本操作。

使用桥接模式，可以让系统在"抽象层次"和"实现层次"这两个层次上独立变化，而又不让系统变得额外复杂。在桥接模式中，让"抽象层次"和"实现层次"是松耦合关系，即"抽象层次"

组合"实现层次"(组合关系),"实现层次"不是"抽象层次"的子类(不是继承关系),从而使两个层次可以相对独立地变化,这就是桥接模式要体现的设计理念。

▶ 16.2.2 结构的描述

下面通过一个简单的问题来描述桥接模式中所涉及的各个角色。

使用桥接模式模拟出版社编辑策划出版 Java 程序设计和 C 程序设计方面的教材。

1. 抽象

抽象角色是 Edit 类。代码如下:

Edit.java

```java
public abstract class Edit {
    String bookName,                                //存放策划信息,例如书名、ISBN、书的内容等
          bookISBN,
          authorName,
          bookContent;
    BookImplementor author;                         //负责基本操作(例如图书内容)
    public abstract void setBookName();             //设置书名
    public abstract void setISBN(String str);       //设置 ISBN
    public abstract void setAuthorName();           //设置作者姓名
    public abstract void setBookContent();          //设置书的内容
    public void setBookImplementor(BookImplementor author){
        this.author = author;
    }
}
```

2. 实现者

实现者是 BookImplementor 接口。代码如下:

BookImplementor.java

```java
public interface BookImplementor {
    public String giveSelfName();                   //给出自己(作者)的名字
    public String giveBookName();                   //给出书名
    public String giveBookContent();                //给出书的内容
}
```

3. 细化抽象

细化抽象角色是 TUPEdit 类,该类的实例模拟清华大学出版社的编辑。代码如下:

TUPEdit.java

```java
public class TUPEdit extends Edit {
    public void setBookName(){                      //设置书名
        bookName = author.giveBookName();
    }
    public void setISBN(String str){                //设置 ISBN
        bookISBN = str;
    }
    public void setAuthorName(){                    //设置作者姓名
        authorName = author.giveSelfName();
```

```java
    }
    public void setBookContent(){              //设置书的内容
        bookContent = author.giveBookContent();
    }
}
```

4. 具体实现者

具体实现者角色有两个类，分别是 GengXY 和 ZhangYP，二者都实现了 BookImplementor 接口，二者的实例分别模拟两位图书作者。代码分别如下：

GengXY.java

```java
public class GengXY implements BookImplementor {
    public String giveSelfName(){              //给出自己(作者)的名字
        return "耿祥义";
    }
    public String giveBookName(){              //给出书名
        return "Java2 实用教程第 6 版 - 微课版";
    }
    public String giveBookContent(){           //给出书的内容
        String mess =
        "类与对象,子类与继承,接口与实现,常用实用类,…泛型与集合框架等内容";
        return mess;
    }
}
```

ZhangYP.java

```java
public class ZhangYP implements BookImplementor {
    public String giveSelfName(){              //给出自己(作者)的名字
        return "张跃平";
    }
    public String giveBookName(){              //给出书名
        return "C 程序设计基础 - 第 2 版";
    }
    public String giveBookContent(){           //给出书的内容
        String mess =
        "运算符与表达式,循环语句,函数的结构与调用,指针…读写文件等内容";
        return mess;
    }
}
```

▶ 16.2.3 模式的使用

前面已经使用桥接模式给出了可以使用的类，这些类就是一个小框架，可以使用这个小框架中的类编写应用程序。

下列应用程序(Application.java)演示了策划图书的过程。程序运行效果如图 16.3 所示。

```
编辑策划的图书:
Java2实用教程第6版-微课版
9787302575443
耿祥义
类与对象,子类与继承,接口与实现,常用实用类...泛型与集合框架等内容
编辑策划的图书:
C程序设计基础-第2版
9787302542919
张跃平
运算符与表达式,循环语句,函数的结构与调用,指针...读写文件等内容
```

图 16.3 程序运行效果

Application.java

```java
public class Application{
    public static void main(String args[]) {
        Edit edit = new TUPEdit();
        BookImplementor author = new GengXY();
        edit.setBookImplementor(author);
        edit.setBookName();                    //设置书名
        edit.setISBN("9787302575443");         //设置 ISBN
        edit.setAuthorName();                  //设置作者姓名
        edit.setBookContent();
        System.out.println("编辑策划的图书:");
        System.out.println(edit.bookName);
        System.out.println(edit.bookISBN);
        System.out.println(edit.authorName);
        System.out.println(edit.bookContent);
        System.out.println("编辑策划的图书:");
        edit.setBookImplementor(new ZhangYP());
        edit.setBookName();                    //设置书名
        edit.setISBN("9787302542919");         //设置 ISBN
        edit.setAuthorName();                  //设置作者姓名
        edit.setBookContent();
        System.out.println(edit.bookName);
        System.out.println(edit.bookISBN);
        System.out.println(edit.authorName);
        System.out.println(edit.bookContent);
    }
}
```

16.3 桥接模式的优点

桥接模式具有以下优点。

- 桥接模式分离了实现与抽象,使实现和抽象可以独立地扩展。当修改实现的代码时,不影响抽象的代码。
- 满足开-闭原则。抽象和实现者处于同层次,使系统可独立地扩展这两个层次。增加新的具体实现者,不需要修改细化抽象;反之,增加新的细化抽象,也不需要修改具体实现。
- 抽象和实现者都可以以继承的方式独立地扩充而互不影响,程序在运行期间可以动态地将一个抽象的子类的实例与一个实现者的子类的实例进行组合。

- 实现者层次代码的修改对抽象层不产生影响，即抽象层的代码不必重新编译。

16.4 应用举例——绘制简单图形

1. 设计要求

当绘制图形时，图形本身有自己的基本数据，例如矩形的位置数据，长、宽等数据。对图形的细化处理就是它的外观，例如在某个坐标系中绘制图形。在不同的比例尺的坐标系中绘制同一个图形，所消耗的资源和对图形的外观感觉是不同的（在比例尺大的坐标系中绘制中国地图，绘制的成本就会加大）。

请使用桥接模式设计一些类，让用户在不同的坐标系中绘制基本图形，例如绘制矩形、椭圆等。

2. 设计实现

1）抽象

抽象角色是 Shape 类。代码如下：

Shape.java

```java
import java.io.File;
import java.io.IOException;
import javax.imageio.ImageIO;
import java.awt.image.BufferedImage;
public abstract class Shape {
    DrawImplementor draw;                          //负责绘制图形
    BufferedImage image;
    public abstract void drawShape();              //绘制
    public abstract void fillShape();              //填充
    public abstract double getArea();
    public void setDrawImplementor( DrawImplementor draw){
         this.draw = draw;
    }
    public void saveShape(String shapeName){
        File file = new File(shapeName + ".bmp");  //保存图形到文件
        try{
            ImageIO.write(image,"bmp",file);
        }
        catch(IOException e) { }
    }
}
```

2）实现者

实现者是 DrawImplementor 接口。代码如下：

DrawImplementor.java

```java
import java.awt.image.BufferedImage;
public interface DrawImplementor {
    public BufferedImage drawRectangle(int x, int y, int width, int height);
    public BufferedImage fillRectangle(int x, int y, int width, int height);
    public BufferedImage drawEllipse(int x, int y, int axisX, int axisY);
    public BufferedImage fillEllipse(int x, int y, int axisX, int axisY);
}
```

3) 细化抽象

细化抽象角色是 Rectangle 和 Ellipse 类。代码分别如下:

Rectangle.java

```java
public class Rectangle extends Shape {           //矩形
    int x,y;                                     //矩形的左上角位置
    int width,height;                            //矩形的宽和高
    public Rectangle(){
        width = 1;
        height = 1;
    }
    public Rectangle(int x,int y,int width,int height){
        this.x = x;
        this.y = y;
        this.width = width ;
        this.height = height ;
    }
    public void drawShape(){                     //绘制
        image = draw.drawRectangle(x,y,width,height);
    }
    public void fillShape(){                     //填充
        image = draw.fillRectangle(x,y,width,height);
    }
    public double getArea(){
        return width * height;
    }
}
```

Ellipse.java

```java
public class Ellipse extends Shape {             //椭圆
    int x,y;                                     //椭圆的圆心
    int axisX,axisY;                             //椭圆的长轴和短轴
    public Ellipse(){
        axisX = 1;
        axisY = 1;
    }
    public Ellipse(int x,int y,int axisX,int axisY){
        this.x = x;
        this.y = y;
        this.axisX = axisX ;
        this.axisY = axisY ;
    }
    public void drawShape(){                     //绘制
        image = draw.drawEllipse(x,y,axisX,axisY);
    }
    public void fillShape(){                     //填充
        image = draw.fillEllipse(x,y,axisX,axisY);
    }
    public double getArea(){
        return Math.PI * axisX * axisY;
    }
}
```

4) 具体实现者

具体实现者是 DrawSimpleShape 类。代码如下：

DrawSimpleShape.java

```java
import java.awt.image.BufferedImage;
import java.awt.*;
public class DrawSimpleShape implements DrawImplementor {
    Graphics g;
    BufferedImage image;
    int imageW = 600, imageH = 360;
    int scale = 1;                              //scale 个像素为一个单位(模拟一个像素的长度)
    public DrawSimpleShape(int scale){
        this.scale = scale;
        image =
        new BufferedImage(imageW,imageH,BufferedImage.TYPE_INT_RGB);
        g = image.getGraphics();
        g.setColor(Color.pink);                 //g 默认用白色绘制图形
    }
    public BufferedImage drawRectangle(int x, int y, int width, int height){
        g.fillRect(0,0,imageW,imageH);          //图形的底色是粉色
        g.setColor(Color.black);
        g.drawRect(x * scale, y * scale, width * scale, height * scale);
        return image;
    }
    public BufferedImage fillRectangle(int x, int y, int width, int height){
        g.fillRect(0,0,imageW,imageH);
        g.setColor(Color.black);
        g.fillRect(x * scale, y * scale, width * scale, height * scale);
        return image;
    }
    public BufferedImage drawEllipse(int x, int y, int axisX, int axisY){
        g.fillRect(0,0,imageW,imageH);
        g.setColor(Color.blue);
        g.drawOval(x * scale - axisX * scale, y * scale - axisY * scale, axisX * scale, axisY * scale);
        return image;
    }
    public BufferedImage fillEllipse(int x, int y, int axisX, int axisY){
        g.fillRect(0,0,imageW,imageH);
        g.setColor(Color.blue);
        g.fillOval(x * scale - axisX * scale, y * scale - axisY * scale, axisX * scale, axisY * scale);
        return image;
    }
}
```

5) 应用程序

前面已经使用桥接模式给出了可以使用的类，这些类是一个小框架，可以使用这个小框架中的类编写应用程序。

下列应用程序（Application.java）绘制了矩形和椭圆，并将绘制的图形保存到磁盘上。程序运行效果如图 16.4 所示。

第16章 桥接模式

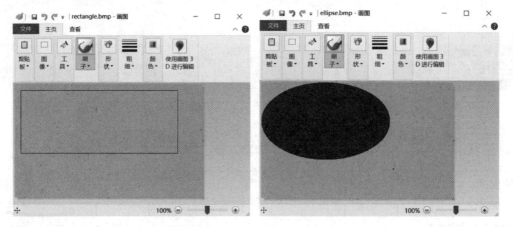

图 16.4 程序运行效果

Application.java

```java
import java.io.*;
public class Application{
    public static void main(String args[]) {
        DrawImplementor implementor = new DrawSimpleShape(10);     //比例尺是 10
        Shape shape = new Rectangle(2,2,50,20);                    //shape 是矩形
        System.out.println("矩形的面积:" + shape.getArea());
        shape.setDrawImplementor(implementor);                     //指定负责绘制图形的实现者
        shape.drawShape();                                         //绘制矩形
        shape.saveShape("rectangle");                              //保存矩形
        Runtime ce = null;
        File file = null;
        try{                                                       //用本地"画笔"程序查看图像
            ce = Runtime.getRuntime();
            file = new File("mspaint.exe rectangle.bmp");
            ce.exec(file.getName());
        }
        catch(Exception e) {}
        implementor = new DrawSimpleShape(20);                     //比例尺是 20
        shape = new Ellipse(20,12,20,12);                          //shape 是椭圆
        shape.setDrawImplementor(implementor);                     //指定负责绘制图形的实现者
        System.out.println("椭圆的面积:" + shape.getArea());
        shape.fillShape();                                         //绘制椭圆
        shape.saveShape("ellipse");                                //保存椭圆
        try{                                                       //用本地"画笔"程序查看图像
            file = new File("mspaint.exe ellipse.bmp");
            ce.exec(file.getName());
        }
        catch(Exception e) {}
    }
}
```

第 17 章　单件模式

以下文本框中的内容引自 GoF 所著 *Design Patterns*：*Elements of Reusable Object Oriented Software* 的中译本及英文版。

> **单件模式**
> 　　保证一个类仅有一个实例，并提供一个访问它的全局访问点。
> **Prototype Pattern**
> 　　Ensure a class only has one instance, and provide a global point of access to it.

以上内容是 GoF 对单件模式的高度概括，结合 17.2.1 节的单件模式的类图可以准确地理解该模式。

17.1　概述

在某些情况下，我们可能需要某个类只能创建一个对象，即不让用户用该类实例化多于一个的实例。例如，在实际生活中，当时间是北京时间 12 点，那么身处北京的人，在晴朗的天气，可以看见天空的太阳，此时，身处纽约的人就看不见天空的太阳，如图 17.1 所示。

| 身处北京的人看天空 | 身处纽约的人看天空 |

图 17.1　唯一的太阳

单件模式是关于怎样设计一个类，并使该类只有一个实例的成熟模式，该模式的关键是将类的构造方法设置为 private（私有）权限，并提供一个返回它的唯一实例的类方法（用关键字 static 修饰的方法）。

视频讲解

17.2　模式的结构与使用

单件模式的结构非常简单，只包括一种角色，即单件类（Singleton）。单件类只可以创建一个实例。

▶ 17.2.1 单件模式的 UML 类图

单件模式的类图如图 17.2 所示。

单件类中含有自身声明的类变量(也称静态变量,即用关键字 static 修饰的成员变量),这个类变量是单件类中唯一的对象,简称"单件对象"。

经常使用以下方法实现一个单件类。

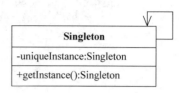

图 17.2 单件模式的类图

(1) 为了确保单件类中自身声明的类变量是单件对象,单件类必须将构造方法的访问权限设置为 private。这样一来,任何其他类都无法使用单件类创建对象(确保了单件类中用自身声明的类变量是单件对象)。

(2) 单件类提供一个类方法(也称静态方法,即用 static 修饰的方法),以便其他用户使用单件类的类名调用这个类方法得到单件对象。

使用以上两种方法实现的单件类如下列 Singleton.java 代码所示:

Singleton.java

```java
public class Singleton {                                    //单件类
    private static Singleton uniqueInstance;                //单件对象
    private Singleton(){}                                   //构造方法是 private 权限
    public static synchronized Singleton getInstance(){     //这是一个同步的类方法
        if(uniqueInstance == null){
            uniqueInstance = new Singleton();               //创建单件对象
        }
        return uniqueInstance;
    }
}
```

▶ 17.2.2 结构的描述

下面通过一个简单的问题来描述单件模式中所涉及的角色。
使用单件模式模拟太阳。

- 单件类

抽象角色是 Sun 类,模拟太阳。代码如下:

Sun.java

```java
import javax.swing.JPanel;
import java.awt.Graphics;
import java.awt.Image;
import java.awt.Toolkit;
public class Sun extends JPanel {                    //单件类(模拟太阳)
    private static Sun sun;                          //单件对象
    private Sun(){}                                  //构造方法是 private 权限
    public static synchronized Sun getInstance(){
        if(sun == null){
            sun = new Sun();                         //创建单件对象
        }
```

```
            return sun;
        }
    public void paint(Graphics g ) {
        Toolkit tool = getToolkit();
        Image img = tool.getImage("sun.jpg");
        g.drawImage(img,0,0,getBounds().width,getBounds().height,this);
    }
}
```

▶ 17.2.3 模式的使用

前面已经使用单件模式给出了可以使用的类,这些类就是一个小框架,可以使用这个小框架中的类编写应用程序。

下列应用程序(Application.java)演示北京时间12点,身处北京的人能看见太阳;纽约时间12点,身处纽约的人能看见太阳。程序运行效果如图17.3所示。

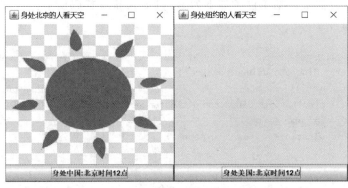

(a) 在北京看太阳

(b) 在纽约看太阳

图 17.3 程序运行效果

Application.java

```
import javax.swing.*;
import java.awt.*;
public class Application{
    public static void main(String args[]){
```

```java
        Sun sunOne = Sun.getInstance();                    //得到单件对象:sun
        Sun sunTwo = Sun.getInstance();                    //得到单件对象:sun
        JFrame Beijing = new JFrame("身处北京的人看天空");
        JFrame NewYork = new JFrame("身处纽约的人看天空");
        JButton beijingButton = new JButton();
        JButton newyorkButton = new JButton();
        NewYork.add(newyorkButton,BorderLayout.SOUTH);
        Beijing.add(beijingButton,BorderLayout.SOUTH);
        Beijing.setBounds(10,10,300,300);
        NewYork.setBounds(320,10,300,300);
        Beijing.setDefaultCloseOperation(JFrame.DISPOSE_ON_CLOSE);
        NewYork.setDefaultCloseOperation(JFrame.DISPOSE_ON_CLOSE);
        Beijing.setVisible(true);
        NewYork.setVisible(true);
        System.out.println("北京时间12点：");
        beijingButton.setText("身处中国:北京时间12点");
        newyorkButton.setText("身处美国:北京时间12点");
        NewYork.add(sunTwo);
        Beijing.add(sunOne);                               //身处北京可见太阳
        Beijing.repaint();
        NewYork.repaint();
        Beijing.validate();
        NewYork.validate();
        try {
            Thread.sleep(5000);
        }
        catch(InterruptedException exp){}
        System.out.println("纽约时间12点：");
        beijingButton.setText("身处中国:纽约时间12点");
        newyorkButton.setText("身处美国:纽约时间12点");
        Beijing.add(sunOne);
        NewYork.add(sunTwo);                               //身处纽约可见太阳
        Beijing.repaint();
        NewYork.repaint();
        Beijing.validate();
        NewYork.validate();
    }
}
```

17.3 单件模式的优点

单件对象由单件类控制，单件类可以很好地控制单件对象。

17.4 应用举例——多线程争冠军

扫一扫

视频讲解

1．设计要求

设计一个单件类以及多个线程。每个线程从左向右水平移动一个属于自己的按钮，最先将按钮移动到指定位置的线程为冠军,即该线程将负责使用单件类得到单件对象(冠军),后续将自己的按钮移动到指定位置的其他线程都可以看到单件对象(冠军)的有关信息。

2. 设计实现

1)单件类

单件类是 Champion 类。代码如下:

Champion.java

```
public class Champion {
    private static Champion first;              //单件对象
    String message;
    private Champion(String message){
        this.message = message;
    }
    public static synchronized Champion getChampion(String s){
        if(first == null){
            first = new Champion(s + "是冠军");
        }
        return first;
    }
    public String getMess(){
        return message;
    }
}
```

2)应用程序

前面已经使用单件模式给出了可以使用的类,这些类就是一个小框架,可以使用这个小框架中的类编写应用程序。

下列应用程序(Application.java)模拟了 500 米赛跑。程序运行效果如图 17.4 所示。

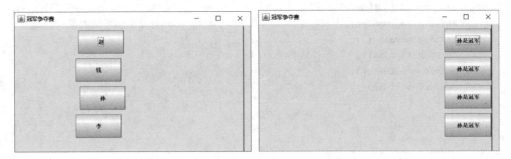

图 17.4 程序运行效果

Application.java

```
import javax.swing.*;
public class Application {
    public static void main(String args[]){
        JFrame win = new JFrame("冠军争夺赛");
        int maxDistance = 500;                  //500 米赛跑
        JButton lineEnd = new JButton();        //终点线
        Target target1 = new Target(maxDistance);
```

第17章 单件模式

```java
            Target target2 = new Target(maxDistance);
            Target target3 = new Target(maxDistance);
            Target target4 = new Target(maxDistance);
            Thread zhao = new Thread(target1);
            Thread qian = new Thread(target2);
            Thread sun = new Thread(target3);
            Thread li = new Thread(target4);
            zhao.setName("赵");
            qian.setName("钱");
            sun.setName("孙");
            li.setName("李");
            win.setLayout(null);
            win.add(target1);
            win.add(target2);
            win.add(target3);
            win.add(target4);
            lineEnd = new JButton();
            win.add(lineEnd);
            target1.setBounds(2,10,100,50);
            target2.setBounds(2,70,100,50);
            target3.setBounds(2,130,100,50);
            target4.setBounds(2,190,100,50);
            lineEnd.setBounds(maxDistance,2,2,700);
            win.setBounds(10,2,700,500);
            win.setDefaultCloseOperation(JFrame.EXIT_ON_CLOSE);
            win.setVisible(true);
            win.validate();
            zhao.start();
            qian.start();
            sun.start();
            li.start();
      }
}
```

Target.java

```java
import javax.swing.JButton;
import java.util.Random;
public class Target extends JButton implements Runnable {
     Champion first;
     int max;
     Random random;
     public Target(int max){
         this.max = max;
         random = new Random();
     }
     public void run(){
         String name = Thread.currentThread().getName();
         setText(name);
         int a = getBounds().x;
         int b = getBounds().y;
```

```java
        int w = getBounds().width;
    while(true) {
        if(a + w >= max){
            first = Champion.getChampion(name);         //获得冠军
            setText(first.getMess());
            return;
        }
        if(random.nextInt(2) == 1)
           a++;
        else
           a = a + 2;
        setLocation(a,b);
        try{
           Thread.sleep(200);
        }
        catch(InterruptedException exp){}
    }
  }
}
```

第 18 章 适配器模式

以下文本框中的内容引自 GoF 所著 *Design Patterns*: *Elements of Reusable Object Oriented Software* 的中译本及英文版。

> **适配器模式（别名：包装器）**
> 将一个类的接口转换成客户希望的另外一个接口。适配器模式使原本由于接口不兼容而不能一起工作的那些类可以一起工作。
> **Adapter Pattern（Another Name：Wrapper）**
> Convert the interface of a class into another interface clients expect. Adapter lets classes work together that couldn't otherwise because of incompatible interfaces.

以上内容是 GoF 对适配器模式的高度概括，结合 18.2.1 节的适配器模式的类图可以准确地理解该模式。

18.1 概述

在实际生活中有很多和适配器类似的问题，例如录音机只能使用直流电，由于供电系统供给用户家里的是交流电，因此用户需要用适配器将交流电转化为直流电，再供录音机使用，如图 18.1 所示。

图 18.1 电源适配器

适配器模式是将一个接口（被适配者）转换成客户希望的另外一个接口（目标）的成熟模式。该模式涉及目标、被适配者和适配器。适配器模式的关键是建立一个适配器，这个适配器实现了目标接口并包含被适配者的引用。

18.2 模式的结构与使用

适配器模式的结构中包括三种角色。

1. 目标（Target）

目标是一个接口，该接口是客户想使用的接口。

2. 被适配者(Adaptee)

被适配者是一个已经存在的接口或抽象类,这个接口或抽象类需要适配。

3. 适配器(Adapter)

适配器是一个类,该类实现了目标接口并包含被适配者的对象,即适配器的职责是对被适配者接口(抽象类)与目标接口进行适配。

18.2.1 适配器模式的UML类图

适配器模式的类图如图18.2所示。

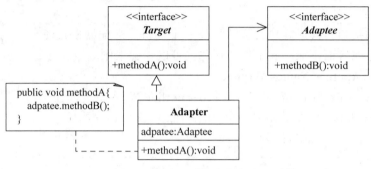

图18.2 适配器模式的类图

18.2.2 结构的描述

下面通过一个简单的问题来描述适配器模式中所涉及的各个角色。

用适配器模式设计几个类,模拟电源适配器将交流电转换为直流电。

1) 目标

目标是一个接口,该接口是客户想使用的接口,本问题中就是用户想使用的直流电。在这里,目标是名字为DirectCurrent的接口。代码如下:

DirectCurrent.java

```
public interface DirectCurrent {         //直流电接口
    public String giveDirectCurrent();   //供给直流电
}
```

注意:实现public String giveDirectCurrent()方法时,可以返回形如1111111111111111的字符序列表示输出直流电。

2) 被适配者

被适配者是一个已经存在的接口或抽象类,这个接口或抽象类需要适配。对于本问题,就是客户已有的交流电。在这里,被适配者是名字为AlternateCurrent的接口。代码如下:

AlternateCurrent.java

```
public interface AlternateCurrent {          //交流电接口
    public String giveAlternateCurrent();    //供给交流电
}
```

> **注意**：在实现 public String giveAlternateCurrent()方法时，可以返回形如 10101010101010101 的字符序列表示输出交流电。

以下的 PowerPlant 类实现了 AlternateCurrent 接口，提供交流电（模拟发电厂）。

PowerPlant.java

```java
public class PowerPlant implements AlternateCurrent {       //交流电提供者
    public String giveAlternateCurrent(){
        return "10101010101010101010";                      //用1、0交替的字符序列表示交流电
    }
}
```

3）适配器

适配器是一个类，该类实现了目标接口并包含被适配者的引用，即适配器的职责是对被适配者接口与目标接口进行适配。在本问题中，适配器的名字是 ElectricAdapter 类，该类实现了 DirectCurrent 接口并包含 AlternateCurrent 接口声明的变量。代码如下：

ElectricAdapter.java

```java
public class ElectricAdapter implements DirectCurrent{
    AlternateCurrent alternateCurrent;               //含有被适配者(交流电接口)
    ElectricAdapter(AlternateCurrent alternateCurrent){
        this.alternateCurrent = alternateCurrent;
    }
    public String giveDirectCurrent(){
        String dc = null;
        System.out.println("这是电源适配器：");
        //从交流电接口(alternateCurrent)得到交流电
        String ac = alternateCurrent.giveAlternateCurrent();
        System.out.println("从交流电接口得到交流电：" + ac);
        StringBuffer strBuffer = new StringBuffer(ac);
        //以下将交流电转为直流电：
        for(int i = 0;i < strBuffer.length();i++) {
            if(strBuffer.charAt(i) == '0') {
                strBuffer.setCharAt(i,'1');
            }
        }
        dc = new String(strBuffer);
        System.out.println("转换为直流电：" + dc);
        return dc;                                    //返回直流电(用字符1组成的字符序列表示直流电)
    }
}
```

18.2.3 模式的使用

前面已经使用适配器模式给出了可以使用的类，这些类就是一个小框架，可以使用这个小框架中的类编写应用程序。

下列应用程序（Application.java）演示了使用电源适配器将交流电转换为直流电供录音机使用。程序运行效果如图 18.3 所示。

```
这是电源适配器：
从交流电接口得到交流电：10101010101010101010
转换为直流电：11111111111111111111
录音机使用直流电：
11111111111111111111
开始录音。
```

图 18.3　程序运行效果

Application.java

```java
public class Application{
    public static void main(String args[]){
        AlternateCurrent alternateCurrent = new PowerPlant();        //发电厂提供交流电
        DirectCurrent directCurrent =                                //直流电
            new ElectricAdapter(alternateCurrent);                   //适配器将交流电适配成直流电
        Recorder recorder = new Recorder();
        recorder.turnOn(directCurrent);                              //录音机使用直流电
    }
}
class Recorder {
    public void turnOn(DirectCurrent DC){                            //录音机使用直流电
        String dc = DC.giveDirectCurrent();
        System.out.println("录音机使用直流电：\n" + dc);
        System.out.println("开始录音。");
    }
}
```

18.3　适配器模式的优点

目标和被适配者是完全解耦的关系。适配器模式满足"开-闭"原则，当添加一个实现被适配者的新类时，不必修改适配器，适配器就能对这个新类的实例进行适配。

18.4　应用举例——替换旧的加密、解密接口

视频讲解

1. 设计要求

目前有一个运行良好的系统，该系统中有负责加密、解密的接口，以及一个实现了该接口的类。但是，当初开发该系统的是一个德国企业，接口中的方法名字都是德语词汇。这个系统除了接口的名字以外，其他功能都很完善。现在决定用新的接口（方法名字都是英语词汇）替代原来的接口，但仍然需要原系统提供的功能。请用适配器模式设计几个类，让用户使用新的接口来运行该系统。

2. 设计实现

1）目标

目标是一个名字为 EncryptionDecryption 的接口，客户想使用这个接口来加密、解密字符序列。代码如下：

EncryptionDecryption.java

```java
public interface EncryptionDecryption {                                    //加密、解密接口
    public String doEncryption(String source,String password);             //加密,返回密文
    public String doDecryption(String secret,String password);             //解密,返回明文
}
```

2）被适配者

被适配是旧系统中的 Secret 接口。代码如下（方法名都是德语词汇）：

Secret.java

```java
public interface Secret {                                                  //加密、解密接口
    public String verschlüsselung(String source,String password);          //加密,返回密文
    public String entschlüsseln(String secret,String password);            //解密,返回明文
}
```

旧系统中已经实现 Secret 接口的 German 类。代码如下：

German.java

```java
import java.util.*;
public class German implements Secret {
    public String verschlüsselung(String source,String password){          //加密,返回密文
        Deque<Character> deque = new ArrayDeque<Character>();              //使用队列加密字符序列
        for(int i = 0;i < password.length();i++){
            deque.add(password.charAt(i));                                 //密码加入队列
        }
        //每次从队头取出一个字符参与加密原文,然后再重新入队,一直到原文被加密完毕
        char a[] = source.toCharArray();
        for(int i = 0;i < a.length;i++){
            char c = deque.pollFirst();                                    //出队操作
            a[i] = (char)(a[i] + c);
            deque.offerLast(c);                                            //c 重新入队
        }
        return new String(a);
    }
    public String entschlüsseln(String secret,String password){            //解密,返回明文
        Deque<Character> deque = new ArrayDeque<Character>();              //使用队列解密字符序列
        for(int i = 0;i < password.length();i++){
            deque.add(password.charAt(i));                                 //密码加入队列
        }
        //每次从队头取出一个字符参与解密,然后再重新入队,一直到密文被解密
        char a[] = secret.toCharArray();
        for(int i = 0;i < a.length;i++){
            char c = deque.pollFirst();                                    //出队操作
            a[i] = (char)(a[i] - c);
            deque.offerLast(c);                                            //c 重新入队
        }
        return new String(a);
    }
}
```

3）适配器

适配器是 GermanAdapter 类。代码如下：

GermanAdapter.java

```java
public class GermanAdapter implements EncryptionDecryption{
    Secret secret;                                         //含有被适配者(旧的德语接口)
    GermanAdapter(Secret secret){
        this.secret = secret;
    }
    public String doEncryption(String source,String password) { //加密,返回密文
        String backStr = secret.verschlüsselung(source,password);
        return backStr;
    }
    public String doDecryption(String secretStr,String password){   //解密,返回明文
        String backStr = secret.entschlüsseln(secretStr,password);
        return backStr;
    }
}
```

4）应用程序

前面已经使用适配器模式给出了可以使用的类，这些类就是一个小框架，可以使用这个小框架中的类编写应用程序。

下列应用程序（Application.java）借助适配器使用新的接口加密、解密字符序列。程序运行效果如图18.4所示。

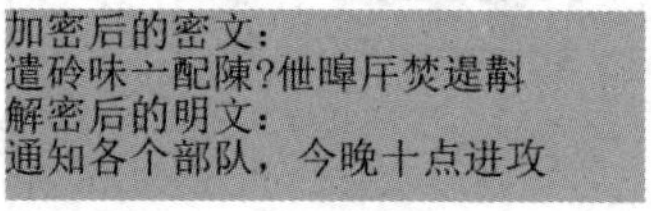

加密后的密文：
遣砖味⌒配陳?伽哖厈焚遑斛
解密后的明文：
通知各个部队，今晚十点进攻

图 18.4　程序运行效果

Application.java

```java
public class Application {
    public static void main(String args[]){
        String str = "通知各个部队,今晚十点进攻";
        String password = "ILoveThisGame";
        EncryptionDecryption ED = new GermanAdapter(new German());   //使用新接口
        String secretStr = ED.doEncryption(str,password);
        System.out.println("加密后的密文:");
        System.out.println(secretStr);
        System.out.println("解密后的明文:");
        String sourceStr = ED.doDecryption(secretStr,password);
        System.out.println(sourceStr);
    }
}
```

第 19 章 模板方法模式

以下文本框中的内容引自 GoF 所著 *Design Patterns*：*Elements of Reusable Object Oriented Software* 的中译本及英文版。

> **模板方法模式**
>
> 定义一个操作中算法的骨架，而将一些步骤延迟到子类中。模板方法使子类可以不改变一个算法的结构即可重定义该算法的某些特定步骤。
>
> **Template Method Pattern**
>
> Define the skeleton of an algorithm in an operation, deferring some steps to subclasses. Template Method lets subclasses redefine certain steps of an algorithm without changing the algorithm's structure.

以上内容是 GoF 对模板方法模式的高度概括，结合 19.2.1 节的模板方法模式的类图可以准确地理解该模式。

19.1 概述

在某些问题的设计中可能遇到这样的现象，一个类和另外一个类中有相同的方法以及执行顺序，但具体的方法内容不尽相同。例如，在乘坐地铁、火车、飞机等交通工具时，都要经历三个基本步骤：①安检，②验票，③乘坐交通工具，如图 19.1 所示。但是，对于不同的交通工具，三个步骤的细节可能有差别，例如，航空的安检就比铁路的更加严格，验票方式也有所不同，提供的交通工具更是不同。

图 19.1　基本步骤

模板方法是关于怎样将若干个方法集成到一个方法中，以便形成一个解决问题的算法骨架。模板方法模式的关键是在一个抽象类中定义一个算法的骨架，即将若干个方法集成到一个方法中，并称该方法为一个模板方法，简称模板。模板方法所调用的其他方法通常为抽象的方法，这些抽象方法相当于算法骨架中的各个步骤，这些步骤的实现可以由子类去完成。

前面提到的三个基本步骤——安检、验票、乘坐交通工具——就是模板方法中的三个方法。航空和铁路部门相当于子类；航空和铁路部门在安检、验票、乘坐交通工具的细节上是不同的。

19.2 模式的结构与使用

模板方法模式包括两种角色。

1. 抽象模板(Abstract Template)

抽象模板是一个抽象类。抽象模板定义了若干个方法以表示一个算法的各个步骤,这些方法中有抽象方法,也有非抽象方法。其中,抽象方法称作原语操作(Primitive Operation)。重要的一点是,抽象模板中还定义了一个称作"模板方法"的方法(非抽象方法),简称模板。模板是算法的骨架,不仅包含原语操作的调用,也可包含非抽象方法的调用。

2. 具体模板(Concrete Template)

具体模板是抽象模板的子类,实现抽象模板中的原语操作。

▶ 19.2.1 模板方法模式的 UML 类图

模板方法模式的类图如图 19.2 所示。

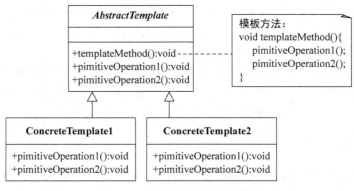

图 19.2 模板方法模式的类图

▶ 19.2.2 结构的描述

下面通过一个简单的问题描述模板方法模式中所涉及的各个角色。

用模板方法模式设计几个类,模拟乘坐不同交通工具所经历的三个基本步骤。

1. 抽象模板

本问题中,抽象模板角色是 Traffic 类。代码如下:

Traffic.java

```java
public abstract class Traffic {
    public void basicStepsTemplate(){          //模板(模板方法)
        securityCheck();                        //步骤①,安检
        if(isTicket()){
            inspectTicket();                    //步骤②,验票
        }
        provideVehicle();                       //步骤③,乘坐交通工具

    }
    public abstract void securityCheck();       //安检(原语操作)
```

```java
    public abstract void inspectTicket();           //验票(原语操作)
    public abstract void provideVehicle();          //乘坐交通工具(原语操作)
    public boolean isTicket(){                      //钩子方法,方法类型是 boolean
        return true;
    }
}
```

2. 具体模板

具体模板有 AircraftTraffic、HighSpeedRailTraffic 和 SpecialTraffic 三个类,分别模拟航空、高铁和特殊交通工具。代码分别如下:

AircraftTraffic.java

```java
public class AircraftTraffic extends Traffic {           //航空
    public void securityCheck(){                         //安检
        System.out.println
        ("安检:禁止携带枪支、爆炸物品类,管制刀具,易燃、易爆物品(不可以携带打火机)……");
    }
    public void inspectTicket(){                         //验票
        System.out.println("验票:出示登机牌");
    }
    public void provideVehicle() {                       //乘坐交通工具
        System.out.println("交通工具:欢迎乘坐民航客机");
    }
}
```

HighSpeedRailTraffic.java

```java
public class HighSpeedRailTraffic extends Traffic {      //高铁
    public void securityCheck(){                         //安检
        System.out.println
        ("安检:禁止携带枪支、爆炸物品类,管制刀具,易燃、易爆物品(可以携带打火机)……");
    }
    public void inspectTicket(){                         //验票
        System.out.println("验票:出示车票、居民身份证、二维码或进行人脸识别");
    }
    public void provideVehicle() {                       //乘坐交通工具
        System.out.println("交通工具:欢迎乘坐中国高铁");
    }
}
```

SpecialTraffic.java

```java
public class SpecialTraffic extends Traffic {            //特殊交通工具
    public void securityCheck(){                         //安检
        System.out.println("安检:\n 禁止携带手机,易燃、易爆物品……");
    }
    public void inspectTicket(){                         //验票
        System.out.println("不需要验票");
    }
}
```

```java
    public void provideVehicle() {              //乘坐交通工具
        System.out.println("交通工具:特殊交通工具.");
    }
    public boolean isTicket(){                  //钩子方法,方法类型是 boolean
        return false;                           //不验票
    }
}
```

注意：钩子方法(返回类型是 boolean 类型)是抽象模板中定义的具体方法,钩子方法的作用是对模板方法中的某些步骤进行"挂钩",即允许具体模板对算法的不同点进行"挂钩",以确定在什么条件下执行模板方法中的哪些算法步骤。具体模板可以根据需要直接继承钩子方法或重写钩子方法。

▶ 19.2.3 模式的使用

前面已经使用模板方法模式给出了可以使用的类,这些类就是一个小框架,可以使用这个小框架中的类编写应用程序。

下列应用程序(Application.java)演示了乘坐不同交通工具的三个基本步骤,但乘坐特殊交通工具省略了一个步骤(验票步骤)。程序运行效果如图 19.3 所示。

```
乘坐航空客机的基本步骤:
安检:禁止携带枪支、爆炸物品类,管制刀具,易燃、易爆物品(不可以携带打火机)……
验票:出示登机牌
交通工具:欢迎乘坐民航客机
乘坐中国高铁的基本步骤:
安检:禁止携带枪支、爆炸物品类,管制刀具,易燃、易爆物品(可以携带打火机)……
验票:出示车票、居民身份证、二维码或进行人脸识别
交通工具:欢迎乘坐中国高铁
乘坐特殊交通工具的基本步骤:
安检:
禁止携带手机,易燃、易爆物品……
交通工具:特殊交通工具
```

图 19.3　程序运行效果

Application.java

```java
public class Application{
    public static void main(String args[]){
        Traffic traffic = new AircraftTraffic();              //航空
        System.out.println("乘坐航空客机的基本步骤:");
        traffic.basicStepsTemplate();
        traffic = new HighSpeedRailTraffic();                 //高铁
        System.out.println("乘坐中国高铁的基本步骤:");
        traffic.basicStepsTemplate();
        traffic = new SpecialTraffic();                       //特殊交通工具
        System.out.println("乘坐特殊交通工具的基本步骤:");
        traffic.basicStepsTemplate();
    }
}
```

19.3　模板方法模式的优点

可以通过在抽象模板的模板方法（模板）中给出成熟的算法步骤，同时又不限制步骤的细节，具体模板实现算法细节不会改变整个算法的骨架。在抽象模板模式中，可以通过钩子方法对某些步骤进行挂钩，具体模板通过钩子可以选择算法骨架中的某些步骤。

19.4　应用举例——数据挖掘

扫一扫

视频讲解

1. 设计要求

挖掘文件中的数据的基本步骤是：①读取文件；②提取数据；③分析数据。目前有一个文件 source.txt，内容如下：

入学时间是 2022-09-03，新生一共 6365 人，其中，男生 3689 人，女生 2676 人。

新生男生平均身高 1.72 米，女生平均身高 1.61 米，总体平均身高 1.66 米。

放假时间是 2023 年 1 月 22 日。

有人想挖掘其中的整型数据（不要日期中的整数）；有人想挖掘浮点型数据；有人想挖掘日期数据。请使用模板方法设计几个类，满足用户的需求。

2. 设计实现

1）抽象模板

抽象模板是 FindData 类。代码如下：

FindData.java

```java
import java.io.File;
public abstract class FindData {
    public void findData(File file) {            //模板方法（模板）
        String dataSource = readFile(file);       //①读取文件
        Object data[] = extractData(dataSource);  //②提取数据
        analysisData(data);                       //③分析数据
    }
    public abstract String readFile(File file);              //原语操作
    public abstract Object[] extractData(String source);     //原语操作
    public abstract void analysisData(Object data[]);        //原语操作
}
```

2）具体模板

具体模板有三个类，分别是 FindFloatData（挖掘浮点数）、FindIntegerData（挖掘整数）和 FindDateData（挖掘日期）。代码分别如下：

FindFloatData.java

```java
import java.util.regex.Pattern;
import java.util.regex.Matcher;
import java.util.Arrays;
import java.util.ArrayList;
import java.io.File;
import java.io.FileInputStream;
import java.io.IOException;
```

```java
public class FindFloatData extends FindData {                    //挖掘浮点数
    public String readFile(File file){
        StringBuffer buffer = new StringBuffer();
        byte [] b = new byte[100];
        String backStr = null;
        try{
            FileInputStream in = new FileInputStream(file);
            int m = -1;
            while((m = in.read(b))!= -1){
                String str = new String(b,0,m);
                buffer.append(str);
            }
            backStr = new String(buffer);
        }
        catch(IOException exp){
            System.out.println("无数据源");
            return null;
        }
        return backStr;
    }
    public Double[] extractData(String source){
        String regex = "-?[0-9][0-9]*[.]+[0-9]*";    //匹配浮点数,不包括整数
        Pattern pattern;                              //模式对象
        Matcher matcher;                              //匹配对象
        pattern = Pattern.compile(regex);             //初始化模式对象
        matcher = pattern.matcher(source);
        ArrayList<String> list = new ArrayList<String>();
        while(matcher.find()) {
            String str = matcher.group();
            list.add(str);
        }
        Double [] number = new Double[list.size()];
        for(int i = 0;i < number.length;i++){
            number[i] = Double.parseDouble(list.get(i));
        }
        return number;
    }
    public void analysisData(Object data[]){
        System.out.println("浮点数据:" + Arrays.toString(data));
    }
}
```

FindIntegerData.java

```java
import java.util.regex.Pattern;
import java.util.regex.Matcher;
import java.util.Arrays;
import java.util.ArrayList;
import java.io.File;
import java.io.FileInputStream;
import java.io.IOException;
```

```java
public class FindIntegerData extends FindData {            //挖掘整数
    public String readFile(File file){
        StringBuffer buffer = new StringBuffer();
        byte [] b = new byte[100];
        String backStr = null;
        try{
            FileInputStream in = new FileInputStream(file);
            int m = -1;
            while((m = in.read(b))!= -1){
                String str = new String(b,0,m);
                buffer.append(str);
            }
            backStr = new String(buffer);
            in.close();
        }
        catch(IOException exp){
            System.out.println("无数据源");
            return null;
        }
        return backStr;
    }
    public Integer[] extractData(String source){
        String regex = "-?[0-9][0-9]*";                    //匹配整数
        Pattern pattern;                                    //模式对象
        Matcher matcher;                                    //匹配对象
        pattern = Pattern.compile(regex);                   //初始化模式对象
        matcher = pattern.matcher(source);
        ArrayList<String> list = new ArrayList<String>();
        while(matcher.find()) {
            String str = matcher.group();
            //返回检索到的字符序列的结束位置后面的位置
            int indexEnd = matcher.end();
            //返回检索到的字符序列的开始位置
            int indexStart = matcher.start();
            boolean isFloat =
                source.charAt(indexEnd) == '.'||
                source.charAt(indexStart >= 1?indexStart - 1:0) == '.';
            boolean isDate =
                source.charAt(indexEnd) == '年'||
                source.charAt(indexEnd) == '月'||
                source.charAt(indexEnd) == '日'||
                source.charAt(indexEnd) == '-'||
                source.charAt(indexEnd) == '/'||
                source.charAt(indexStart - 1) == '/'||
                source.charAt(indexStart) == '-' &&
                source.charAt(indexEnd) == '-'||
                source.charAt(indexStart) == '-' &&
                Character.isDigit(source.charAt(indexStart - 1)) ;
            if(!isDate&&!isFloat) {
                list.add(str);                              //不要日期和浮点数中的数字
            }
```

```java
            }
            Integer [] number = new Integer[list.size()];
            for(int i = 0;i < number.length;i++){
                number[i] = Integer.parseInt(list.get(i));
            }
            return number;
        }
        public void analysisData(Object data[]){
            System.out.println("整型数据:" + Arrays.toString(data));
            System.out.println("排序:");
            Arrays.sort(data);
            System.out.println(Arrays.toString(data));
            System.out.println("数据之间的差");
            for( int i = 0;i < data.length - 1;i++) {
                String s2 = data[i + 1].toString();
                String s1 = data[i].toString();
                System.out.println (Integer.parseInt(s2) - Integer.parseInt(s1));
            }
        }
    }
}
```

FindDateData.java

```java
import java.util.regex.Pattern;
import java.util.regex.Matcher;
import java.util.Arrays;
import java.util.ArrayList;
import java.util.StringTokenizer;
import java.io.File;
import java.io.FileInputStream;
import java.io.IOException;
import java.time.LocalDate;
import java.time.temporal.ChronoUnit;
public class FindDateData extends FindData {            //挖掘日期数
    public String readFile(File file){
        StringBuffer buffer = new StringBuffer();
        byte [] b = new byte[100];
        String backStr = null;
        try{
            FileInputStream in = new FileInputStream(file);
            int m = -1;
            while((m = in.read(b))!= -1){
                String str = new String(b,0,m);
                buffer.append(str);
            }
            backStr = new String(buffer);
            in.close();
        }
        catch(IOException exp){
            System.out.println("无数据源");
            return null;
```

```java
            }
            return backStr;
        }
        public String[] extractData(String source){
            String year = "[1-9][0-9]{3}";
            String month = "((0?[1-9])|(1[012]))";
            String day = "((0?[1-9][^0-9])|([12][0-9])|(3[01]))";
            String regex = year+"[-年/]"+month+"[-月/]"+day;
            Pattern pattern;                        //模式对象
            Matcher matcher;                        //匹配对象
            pattern = Pattern.compile(regex);       //初始化模式对象
            matcher = pattern.matcher(source);
            ArrayList<String> list = new ArrayList<String>();
            while(matcher.find()) {
                String str = matcher.group();
                list.add(str);
            }
            String [] str = new String[list.size()];
            for(int i = 0;i<str.length;i++){
                str[i] = list.get(i);
            }
            return str;
        }
        public void analysisData(Object data[]){
            System.out.println("日期数据:"+Arrays.toString(data));
            System.out.println("各个日期之间的差(按天):");
            for(int i = 0;i<data.length-1;i++) {
                long difference =
                changeDate(data[i]).until(changeDate(data[i+1]),ChronoUnit.DAYS);
                System.out.println(difference);
            }
        }
        LocalDate changeDate(Object object) {
            String str = object.toString();
            StringTokenizer token = new StringTokenizer(str,"-/年月日");
            int year = Integer.parseInt(token.nextToken());
            int month = Integer.parseInt(token.nextToken());
            int day = Integer.parseInt(token.nextToken());
            LocalDate date = LocalDate.of(year,month,day);
            return date;
        }
    }
```

3）应用程序

前面已经使用模板方法模式给出了可以使用的类，这些类就是一个小框架，可以使用这个小框架中的类编写应用程序。

下列应用程序（Application.java）使用模板方法模式中的模板挖掘文件 source.txt 中的数据。程序运行效果如图 19.4 所示。

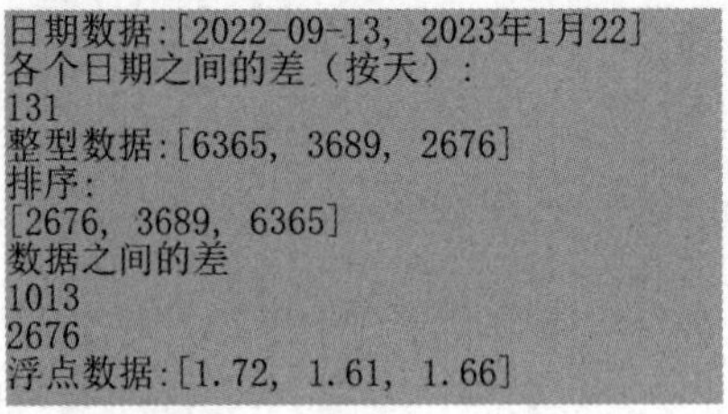

```
日期数据:[2022-09-13, 2023年1月22]
各个日期之间的差(按天):
131
整型数据:[6365, 3689, 2676]
排序:
[2676, 3689, 6365]
数据之间的差
1013
2676
浮点数据:[1.72, 1.61, 1.66]
```

图 19.4 程序运行效果

Application.java

```java
import java.io.File;
public class Application{
    public static void main(String args[]){
        File file = new File("source.txt");
        FindData findData = new FindDateData();           //挖掘日期
        findData.findData(file);
        findData = new FindIntegerData();                 //挖掘整数
        findData.findData(file);
        findData = new FindFloatData();                   //挖掘浮点数
        findData.findData(file);
    }
}
```

第 20 章 外观模式

以下文本框中的内容引自 GoF 所著 *Design Patterns：Elements of Reusable Object Oriented Software* 的中译本及英文版。

> **外观模式**
> 为系统中的一组接口提供一个一致的界面。外观模式定义了一个高层接口,这个接口使这一子系统更加容易使用。
>
> **The Facade Pattern**
> Provide a unified interface to a set of interfaces in a subsystem. Facade defines a higher level interface that makes the subsystem easier to use.

以上内容是 GoF 对外观模式的高度概括,结合 20.2.1 节的外观模式的类图可以准确地理解该模式。

20.1 概述

一个大的系统一般都由若干子系统构成,每个子系统包含多个类,这些类协同合作为用户提供所需要的功能。从需求角度考虑,不能让客户直接和子系统打交道,原因是客户可能不清楚子系统的结构,即不清楚怎样和子系统打交道。另外,从软件设计角度考虑,如果让客户程序直接和子系统的多个类的实例打交道完成某项任务,就使客户程序中的类和子系统类有过多的依赖关系。例如,广告公司有审核、计费和设计三个子系统,一个客户可能不清楚广告公司的这三个子系统,他直接把自己要宣传的产品和需求告诉广告公司的客服(负责接待客户的人员),客服就会根据工作业务流程安排有关工作,最终完成客户的需求,如图 20.1 所示。

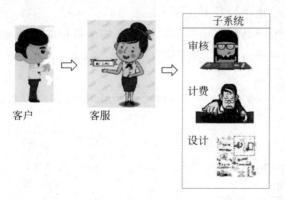

图 20.1 客户与广告公司的子系统

外观模式是简化用户和子系统进行交互的成熟模式。外观模式的关键是为子系统提供一个称作外观的类,该外观类的实例负责和子系统中的类的实例打交道。当用户想要和子系统中的若干类的实例打交道时,外观类的实例可以代替用户和子系统的外观类的实例打交道。例如,前面列举的审核、计费和设计就是广告公司的三个子系统;"客服"就是外观类的实例。

扫一扫

视频讲解

20.2 模式的结构与使用

外观模式包括两种角色。

1. 子系统(Subsystem)

子系统是若干类的集合,这些类的实例协同合作为用户提供所需要的功能,子系统中的任何类都不包含外观类的对象。

2. 外观(Facade)

外观是一个类,该类包含子系统中全部或部分类的对象,当用户想要和子系统中的若干类的实例打交道时,它可以代替用户和子系统的外观类的对象打交道。

▶ 20.2.1 外观模式的 UML 类图

外观模式的类图如图 20.2 所示。

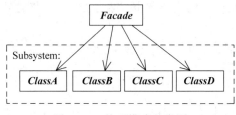

图 20.2 外观模式的类图

▶ 20.2.2 结构的描述

下面通过一个简单的子系统讲述外观模式中所涉及的各个角色。

广告公司的子系统有三个类:Check、Charge 和 TypeSetting。Check 类的对象负责审核广告内容;Charge 类的对象负责计算费用;TypeSetting 类的对象负责排版。

使用外观模式可以简化用户和上述子系统所进行的交互。例如,一个用户想要做广告,那么用户只需要将要宣传的产品和需求告诉广告公司的子系统的外观对象即可,外观对象将负责和子系统中的类的对象进行交互,完成用户所指派的任务。

1. 子系统

子系统有三个类:Check、Charge 和 TypeSetting。代码分别如下:

Check. java

```java
public class Check {
    public boolean examine(String advertisement){
        boolean isOK = true;
        if(advertisement.length()> 300){          //字数显示
            isOK = false;
        }
        if(advertisement.contains("香烟")){        //内容检查
            isOK = false;
```

```
            }
            return isOK;
        }
}
```

Charge.java

```java
import javax.swing.JOptionPane;
public class Charge{
    public final int basicCharge = 12;
    public final int complexCharge = 58;
    public void giveCharge(String advertisement,int imageAmount){
        int charge = advertisement.length() * basicCharge +
                       imageAmount * complexCharge ;
        JOptionPane.showMessageDialog(null,"广告费用:" + charge + "元","消息对话框",
                                JOptionPane.ERROR_MESSAGE);
    }
}
```

TypeSetting.java

```java
import javax.swing.*;
import java.awt.*;
public class TypeSetting extends JFrame {
    Image image;
    JTextArea showWord;
    public void typeSetting(String advertisement){
        showWord = new JTextArea();
        showWord.setText(advertisement);
        showWord.setFont(new Font("",Font.BOLD,26));
        add(showWord,BorderLayout.SOUTH);
        validate();
        Toolkit toolkit = Toolkit.getDefaultToolkit();
        setDefaultCloseOperation(JFrame.EXIT_ON_CLOSE);
        setBounds(12,12,500,600);
        String imageName = JOptionPane.showInputDialog
        (this,"输入图像名称,例如 moon.jpg","输入对话框",JOptionPane.PLAIN_MESSAGE);
        if(imageName!= null){
            image = toolkit.getImage(imageName);
        }
        setVisible(true);
        repaint();
    }
    public void paint(Graphics g ) {
        super.paint(g);
        g.drawImage(image,0,0,getBounds().width,getBounds().height,this);
        showWord.repaint();
    }
}
```

2. 外观

外观是 ServerFacade 类,该类含有 Check、Charge 和 TypeSetting 的对象。ServerFacade

类的代码如下：

ServerFacade.java

```java
import javax.swing.JOptionPane;
public class ServerFacade {
    private Check check;                         //审核
    private Charge charge;                       //计费
    private TypeSetting typeSetting;             //排版
    String advertisement;
    public ServerFacade(){
        check = new Check();
        charge = new Charge();
        typeSetting = new TypeSetting();
    }
    public void receiveAdvertisement(String advertisement,int imageAmount){
        if(check.examine(advertisement)){
            charge.giveCharge(advertisement,imageAmount);
            typeSetting.typeSetting(advertisement);
        }
        else {
            JOptionPane.showMessageDialog(null,"广告审核未通过","消息对话框",
                                JOptionPane.ERROR_MESSAGE);
        }
    }
}
```

▶ 20.2.3 模式的使用

前面已经使用外观模式给出了可以使用的类，这些类就是一个小框架，可以使用这个小框架中的类编写应用程序。

下列应用程序(Application.java)演示了用户将广告交给外观模式中的外观对象。程序运行效果如图 20.3 所示。

图 20.3 程序运行效果

Application.java

```java
public class Application{
    public static void main(String args[]){
        ServerFacade serverFacade;              //外观的实例
        String advertisement = "华为,不仅仅是世界 500 强!细节会让你感动!";
        serverFacade = new ServerFacade();
        serverFacade.receiveAdvertisement(advertisement,1);
    }
}
```

20.3 外观模式的优点

外观模式使客户和子系统中的类无耦合,并且使子系统使用起来更加方便。外观只是提供了一个更加简洁的界面,并不影响用户直接使用子系统中的类。子系统中任何类对其方法的内容进行修改,不影响外观类的代码。

20.4 应用举例——解析文件

扫一扫

视频讲解

1. 设计要求

设计一个子系统,该子系统有三个类:ReadFile、AnalyzeInformation 和 SaveFile,各个类的职责如下:

- ReadFile 类的实例可以读取文本文件。
- AnalyzeInformation 类的实例可以从一个文本中删除用户不需要的内容。
- SaveFile 类的实例可以将一个文本保存到文本文件中。

请为上述子系统设计一个外观,以便简化用户和上述子系统所进行的交互。例如,一个用户想要读取一个 HTML 文件,并将该文件的内容中的全部 HTML 标记去掉后保存到另一个文本文件中,那么用户只需要把要读取的 HTML 文件名、一个正则表达式(表示删除的信息)以及要保存的文件名告诉子系统的外观对象即可,外观对象和子系统中类的对象进行交互,完成用户所指派的任务。

2. 设计实现

1) 子系统

子系统中有三个类:ReadFile、AnalyzeInformation 和 WriteFile。代码分别如下:

ReadFile.java

```java
import java.io.*;
public class ReadFile{
    public byte[] readFileContent(String fileName){
        int n = -1;
        File f = new File(fileName);
        byte [] a = new byte[(int)f.length()];
        try{
            InputStream in = new FileInputStream(f);
            n = in.read(a);
```

```
            in.close();
        }
        catch(IOException e) {}
        return a;
    }
}
```

AnalyzeInformation.java

```
import java.util.regex.*;
public class AnalyzeInformation{
    public String getSavedContent(String content,String deleteContent){
        Pattern p;
        Matcher m;
        p = Pattern.compile(deleteContent);
        m = p.matcher(content);                    //m 可以在 content 中找到 deleteContent
        String savedContent = m.replaceAll("");    //替换掉 deleteContent
        return savedContent;
    }
}
```

WriteFile.java

```
import java.io.*;
public class WriteFile{
    public void writeToFile(String fileName,String content){
        byte [] a = content.getBytes();
        File file = new File(fileName);
        try{
            OutputStream out = new FileOutputStream(file);
            out.write(a);
            out.close();
        }
        catch(IOException e) {}
    }
}
```

2）外观

外观是 Facade 类，该类的对象含有 ReadFile、AnalyzeInformation 和 SaveFile 的对象。Facade 类的代码如下：

Facade.java

```
public class Facade{
    private ReadFile readFile;
    private AnalyzeInformation analyzeInformation;
    private WriteFile writeFile;
    public Facade(){
        readFile = new ReadFile();
        analyzeInformation = new AnalyzeInformation();
        writeFile = new WriteFile();
    }
```

```java
public void doing(String readFileName,String delContent,String savedFileName){
    byte [] a = readFile.readFileContent(readFileName);
    System.out.println("读取文件" + readFileName + "的内容:");
    System.out.println(new String(a));
    String savedContent =
    analyzeInformation.getSavedContent(new String(a),delContent);
    writeFile.writeToFile(savedFileName,savedContent);
    System.out.println("保存到文件" + savedFileName + "中的内容:");
    System.out.println(savedContent);
    }
}
```

3）应用程序

前面已经使用外观模式给出了可以使用的类，这些类就是一个小框架，可以使用这个小框架中的类编写应用程序。

下列应用程序（Application.java）使用了外观模式中所涉及的类，应用程序负责创建外观类的实例。一个用户想要读取一个 HTML 文件，并将该文件内容中的全部 HTML 标记去掉后保存到另一个文本文件中，那么用户只需要把要读取的 HTML 文件名、一个正则表达式（表示删除的信息）以及要保存的文件名告诉子系统 Facade 外观对象即可。程序运行效果如图 20.4 所示。

```
读取文件 index.html 的内容:
<html>清华大学校园网
<center>
<h1> 工程学院</h1>
  <p>欢迎新同学</p>
<h1> 计算机科学技术学院</h1>
  <p>欢迎新同学</p>
</center>
</hml>
保存到文件 save.txt 中的内容:
清华大学校园网

 工程学院
  欢迎新同学
 计算机科学技术学院
  欢迎新同学
```

图 20.4　程序运行效果

Application.java

```java
public class Application{
    public static void main(String args[]){
        Facade serverFacade = new Facade();              //外观
        String readFileName = "index.html";
        String delContent = "<[^>]*>";                   //匹配超文本标记
        String savedFileName = "save.txt";
        serverFacade.doing(readFileName,delContent,savedFileName);
    }
}
```

第 21 章　中介者模式

以下文本框中的内容引自 GoF 所著 *Design Patterns*：*Elements of Reusable Object Oriented Software* 的中译本及英文版。

> **中介者模式**
> 　　用一个中介对象来封装一系列的对象交互。中介者使各对象不需要显式地相互引用，从而使其耦合松散，而且可以独立地改变它们之间的交互。
> **Mediator Pattern**
> 　　Define an object that encapsulates how a set of objects interact. Mediator promotes loose coupling by keeping objects from referring to each other explicitly, and it lets you vary their interaction independently.

　　以上内容是 GoF 对中介者模式的高度概括，结合 21.2.1 节的中介者模式的类图可以准确地理解该模式。

21.1　概述

　　一个对象组合另一个对象是面向对象中经常使用的方式，也是面向对象所提倡的，即"少用继承，多用组合"。但是怎样合理地组合对象对系统今后的扩展、维护和对象的复用是至关重要的，这也正是学习设计模式的重要原因。在面向对象编程中，如果对象 A 组合了对象 B（对象 A 含有对象 B），人们习惯地称 B 是 A 的朋友。如果 B 是 A 的朋友，那么对象 A 就可以请求 B 执行相关的操作。但是，对某些特殊系统，特别是涉及很多对象时，该系统可能不希望这些对象直接交互，即不希望这些对象之间互相包含对方成为朋友，其原因是不利于系统今后的扩展、维护以及对象的复用。例如，在一个房屋租赁系统中有很多对象，有些对象是求租者，有些对象是出租者，如果要求他们之间必须互相成为朋友才能进行有关租赁操作，显然不利于系统的维护和扩展。一个好的解决办法是在房屋租赁系统中建立一个称作中介者的对象，中介者包含系统中的所有其他对象，而系统中的其他对象只包含中介者，也就是说中介者和大家互为朋友。中介者使系统中的其他对象之间完全解耦，当系统中某个对象需要和系统中另外一个对象交互时，只需要将自己的请求通知中介者即可，如图 21.1 所示。

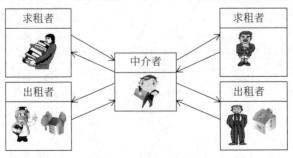

图 21.1　中介者与求租者、出租者

第21章 中介者模式

中介者模式是封装一系列的对象交互的成熟模式,其关键是将对象之间的交互封装在称作中介者的对象中,中介者使各对象不需要显式地相互包含,这些对象只包含中介者。

21.2 模式的结构与使用

中介者模式的结构中包括四种角色。

1. 中介者(Mediator)

中介者是一个接口,该接口定义了用于同事(Colleague)对象之间进行通信的方法。

2. 具体中介者(ConcreteMediator)

具体中介者是实现中介者接口的类。具体中介者需要包含所有具体同事(即具体中介者需要组合所有具体同事)(ConcreteColleague),并通过实现中介者接口中的方法来满足具体同事之间的通信请求。

3. 同事(Colleague)

同事是一个接口,规定了具体同事需要实现的方法。

4. 具体同事(ConcreteColleague)

具体同事是实现同事接口的类。具体同事需要包含具体中介者,一个具体同事需要和其他具体同事交互时,只需要将自己的请求通知给它所包含的具体中介者即可。

21.2.1 中介者模式的 UML 类图

中介者模式的类图如图 21.2 所示。

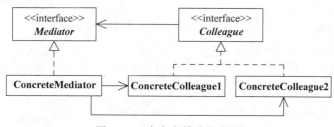

图 21.2 中介者模式的类图

21.1 节提到的房屋租赁系统中的"求租者"和"出租者"都是具体同事。

21.2.2 结构的描述

下面通过一个简单的问题来描述中介者模式中所涉及的各个角色。

目前有一个房屋出租者和两个求租者,他们想通过一个中介者交互有关信息。出租者想给其中一位求租者的信息是"每年 1 万 2 千元,有线电视费由求租者承担。"给另一位求租者的信息是"每年 1 万 5 千元,有线电视费由求租者承担。"一位求租者想给出租者的信息是"希望每 6 个月结算一次租金。"另一位求租者想给出租者的信息是"希望每 2 个月结算一次租金。"

针对上述问题,使用中介者模式设计若干类,模拟中介者的作用。

1. 中介者

中介者是 Mediator 接口。代码如下:

Mediator.java

```java
public interface Mediator {
    public void registerColleague(Colleague colleague);
    public void deliverMess(Colleague colleague,String mess);
}
```

2. 具体中介者

具体中介者是 HouseMediator 类。代码如下：

HouseMediator.java

```java
import java.util.ArrayList;
public class HouseMediator implements Mediator {
    ArrayList<Colleague> list;                    //存放同事的数组表
    public HouseMediator(){
        list = new ArrayList<Colleague>();
    }
    public void registerColleague(Colleague colleague){
        list.add(colleague);                      //把同事添加到数组表
    }
    public void deliverMess(Colleague colleague,String mess){
        Colleague wantColleague = null;           //需要接收消息的同事
        for(Colleague c:list){
            if(c.equals(colleague))
                wantColleague = c;                //找到需要接收消息的同事
        }
        if(wantColleague!= null)
            wantColleague.receiveMess(mess);      //同事接收消息
    }
}
```

3. 同事

同事是 Colleague 接口。代码如下：

Colleague.java

```java
public interface Colleague {
    public void giveMess(Colleague c,String mess);
    public void receiveMess(String mess);
    public void setName(String name);
    public String getName();
}
```

4. 同事

具体同事是 Person 类。代码如下：

Person.java

```java
public class Person implements Colleague {
    Mediator mediator;                            //中介者
    String name;
    public Person(Mediator mediator){
        this.mediator = mediator;
```

```java
            mediator.registerColleague(this);        // 将当前对象注册到中介者
    }
    public void giveMess(Colleague c,String mess){
        mediator.deliverMess(c,mess);                //请中介者转达信息
    }
    public void receiveMess(String mess){
        System.out.println(name + "收到信息:" + mess);
    }
    public void setName(String name){
        this.name = name;
    }
    public String getName(){
        return name;
    }
}
```

▶ 21.2.3 模式的使用

前面已经使用中介者模式给出了可以使用的类,这些类就是一个小框架,可以使用这个小框架中的类编写应用程序。

下列应用程序(Application.java)使用了中介者模式中所涉及的类。求租者和出租者把信息交给中介者,由中介者转达消息。程序运行效果如图 21.3 所示。

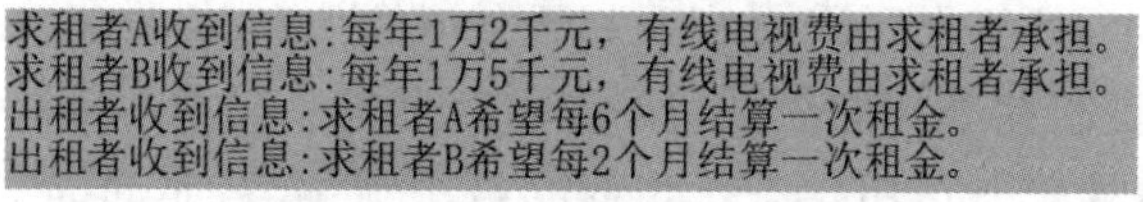

```
求租者A收到信息:每年1万2千元,有线电视费由求租者承担。
求租者B收到信息:每年1万5千元,有线电视费由求租者承担。
出租者收到信息:求租者A希望每6个月结算一次租金。
出租者收到信息:求租者B希望每2个月结算一次租金。
```

图 21.3　程序运行效果

Application.java

```java
public class Application{
    public static void main(String args[]){
        Mediator mediator = new HouseMediator();        //中介者
        Colleague lessor = new Person(mediator);        //出租者
        lessor.setName("出租者");
        Colleague renterA = new Person(mediator);       //求租者 A
        renterA.setName("求租者 A");
        Colleague renterB = new Person(mediator);       //求租者 B
        renterB.setName("求租者 B");
        lessor.giveMess(renterA,"每年 1 万 2 千元,有线电视费由求租者承担。");
        lessor.giveMess(renterB,"每年 1 万 5 千元,有线电视费由求租者承担。");
        renterA.giveMess(lessor,renterA.getName() + "希望每 6 个月结算一次租金。");
        renterB.giveMess(lessor,renterB.getName() + "希望每 2 个月结算一次租金。");
    }
}
```

21.3　中介者模式的优点

中介者模式具有以下优点:
(1)可以避免许多对象相互组合,这有利于系统的维护,也使其他系统可以复用这些

对象。

（2）可以通过中介者将原本分布于多个对象之间的交互行为集中在一起。当需要改变这些对象之间的通信行为时，只需要使用一个具体中介者即可，不必修改各个具体同事的代码，即这些同事可被复用。

（3）具体中介者使各个具体同事完全解耦，修改任何一个具体同事的代码不会影响到其他同事。

（4）具体中介者集中了同事之间交互的细节，使系统比较清楚地知道整个系统中的同事是如何交互的。

（5）当一些对象想互相通信，但又无法相互包含对方时，使用中介者模式可以使这些对象互相通信。

注意：由于具体中介者集中了同事之间交互的细节，可能使具体中介者变得非常复杂，增加维护的难度。

扫一扫

视频讲解

21.4 应用举例——协调复制、剪切与粘贴

1. 设计要求

在 GUI 程序中一个常用的操作是将文本复制或剪切到剪贴板，以及将剪贴板中的文本粘贴到程序中。例如，一个 GUI 程序需要实现如下的功能：程序中有一个文本区，当文本区中有文本被选中时，负责复制和剪切的组件将处于可用状态；当文本区中没有文本被选中时，负责复制和剪切的组件将处于非可用状态。当剪贴板中无内容时，负责粘贴的组件处于非可用状态；当剪贴板中有内容时，负责粘贴的组件处于可用状态。

请使用中介者模式设计几个类，实现复制、剪切与粘贴功能。

2. 设计实现

为实现上述功能，并不需要明确地定义模式中的同事接口和中介者接口，只需要给出具体同事和具体中介者即可。

1) 具体同事

具体同事是 javax.swing 包中的 JMenu 类、JMenuItem 类以及 JTextArea 类。

2) 具体中介者

具体中介者是 ConcreteMediator 类。代码如下：

ConcreteMediator.java

```
import javax.swing.*;
import java.awt.datatransfer.*;
public class ConcreteMediator{
    JMenu menu;
    JMenuItem copyItem,cutItem,pasteItem;
    JTextArea text;
    public void openMenu(){
        Clipboard clipboard = text.getToolkit().getSystemClipboard();
        String str = text.getSelectedText();
        if(str == null){
            copyItem.setEnabled(false);
```

```java
            cutItem.setEnabled(false);
        }
        else{
            copyItem.setEnabled(true);
            cutItem.setEnabled(true);
        }
        boolean boo = clipboard.isDataFlavorAvailable(DataFlavor.stringFlavor);
        if(boo){
            pasteItem.setEnabled(true);
        }
    }
    public void paste(){
        text.paste();
    }
    public void copy(){
        text.copy();
    }
    public void cut(){
        text.cut();
    }
    public void registerMenu(JMenu menu){
        this.menu = menu;
    }
    public void registerPasteItem(JMenuItem item){
        pasteItem = item;
    }
    public void registerCopyItem(JMenuItem item){
        copyItem = item;
        copyItem.setEnabled(false);
    }
    public void registerCutItem(JMenuItem item){
        cutItem = item;
        cutItem.setEnabled(false);
    }
    public void registerText(JTextArea text){
        this.text = text;
    }
}
```

3）应用程序

前面已经使用中介者模式给出了可以使用的类，这些类就是一个小框架，可以使用这个小框架中的类编写应用程序。

下列应用程序（Application.java）使用了中介者模式中所涉及的类，演示复制、剪切和粘贴。程序运行效果如图 21.4 所示。

图 21.4　程序运行效果

Application.java

```java
import javax.swing.*;
import java.awt.event.*;
```

```java
import java.awt.*;
import javax.swing.event.*;
public class Application extends JFrame{
    ConcreteMediator mediator;
    JMenuBar bar;
    JMenu menu;
    JMenuItem copyItem,cutItem,pasteItem;
    JTextArea text;
    public Application(){
        mediator = new ConcreteMediator();
        bar = new JMenuBar();
        menu = new JMenu("编辑");
        menu.setFont(new Font("",Font.BOLD,22));
        menu.addMenuListener(new MenuListener(){
                public void menuSelected(MenuEvent e){
                    mediator.openMenu();
                }
                public void menuDeselected(MenuEvent e){}
                public void menuCanceled(MenuEvent e){}
            });
        copyItem = new JMenuItem("复制");
        copyItem.setFont(new Font("",Font.BOLD,22));
        copyItem.addActionListener((e) -> { mediator.copy();});
        cutItem = new JMenuItem("剪切");
        cutItem.setFont(new Font("",Font.BOLD,22));
        cutItem.addActionListener((e) -> { mediator.cut();});
        pasteItem = new JMenuItem("粘贴");
        pasteItem.setFont(new Font("",Font.BOLD,22));
        pasteItem.addActionListener((e) ->{ mediator.paste();});
        text = new JTextArea();
        text.setFont(new Font("",Font.BOLD,26));
        bar.add(menu);
        menu.add(cutItem);
        menu.add(copyItem);
        menu.add(pasteItem);
        setJMenuBar(bar);
        add(text,BorderLayout.CENTER);
        register();
        setDefaultCloseOperation(JFrame.EXIT_ON_CLOSE);
    }
    private void register(){
        mediator.registerMenu(menu);
        mediator.registerCopyItem(copyItem);
        mediator.registerCutItem(cutItem);
        mediator.registerPasteItem(pasteItem);
        mediator.registerText(text);
    }
    public static void main(String args[]){
        Application application = new Application();
        application.setBounds(100,200,500,360);
        application.setVisible(true);
    }
}
```

第 22 章　迭代器模式

以下文本框中的内容引自 GoF 所著 *Design Patterns：Elements of Reusable Object Oriented Software* 的中译本及英文版。

> **迭代器模式（别名：游标）**
> 　　提供一种方法顺序访问一个聚合对象中的各个元素，而又不需要暴露该对象的内部表示。
>
> **Iterator Pattern（Another Name：Cursor）**
> 　　Provide a way to access the elements of an aggregate object sequentially without exposing its underlying representation.

以上内容是 GoF 对迭代器模式的高度概括，结合 22.2.1 节的迭代器模式的类图可以准确地理解该模式。

22.1　概述

合理组织数据的结构以及相关操作是程序设计的一个重要方面，例如在程序设计中经常会使用诸如链表、散列表等数据结构。链表和散列表等数据结构都可以存放若干个对象的集合，其区别是按照不同的方式来存储对象。用户希望无论何种集合，都允许程序以一种统一的方式遍历集合中的对象，而不需要知道这些对象在集合中是如何表示及存储的。例如，一栋楼中居住着张三、李四、刘五三个人，分别住在不同的房间，张三知道李四的房间，李四知道刘五的房间。假设有一个警察，他并不想知道这些人是以什么方式在此居住，只想找到他们，那么这个警察可以使用一个名字为 next()的方法找人，该方法的特点是在找到一个人的同时立刻让这个人说出他所知道的下一个人所在的房间，然后警察再调用 next()方法找到下一个人。如果警察调用 next()方法首先找到张三，就可以依次地调用 next()方法找到所有人，如图 22.1 所示。

图 22.1　警察找人

迭代器模式是遍历集合的成熟模式,迭代器模式的关键是将遍历集合的任务交给一个称作迭代器的对象。例如,前面所述的楼就是一个集合,其中的人就是集合中的对象,而警察就是一个迭代器。

22.2 模式的结构与使用

迭代器模式的结构中包括四种角色。

1. 集合(Aggregate)

集合角色是一个接口,规定了具体集合需要实现的操作。

2. 具体集合(Concrete Aggregate)

具体集合角色是实现集合接口的类,具体集合按照一定结构存储对象。具体集合有一个方法,该方法返回一个针对该集合的具体迭代器。

3. 迭代器(Iterator)

迭代器角色是一个接口,规定了遍历具体集合的方法,例如 next()方法。

4. 具体迭代器(Concrete Iterator)

具体迭代器角色是实现迭代器接口的类。具体迭代器在实现迭代器接口所规定的遍历集合的方法,例如 next()方法时,要保证 next()方法的首次调用按照集合的数据结构找到该集合中的一个对象,而且每当找到集合中的一个对象时,立刻根据该集合的存储结构得到待遍历的后继对象的引用,并保证依次调用 next()方法可以遍历集合。

▶ 22.2.1 迭代器模式的 UML 类图

迭代器模式的类图如图 22.2 所示。

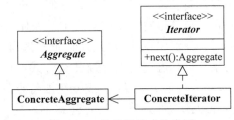

图 22.2 迭代器模式的类图

▶ 22.2.2 结构的描述

下面通过一个简单的问题讲述迭代器模式中所涉及的各个角色。

用一个集合模拟一栋楼,该集合返回的迭代器模拟警察在楼中找人。

1. 集合

集合是名字为 Set 的接口,用到的数据类型是 People(即集合中的元素)。Set 接口和 People 类的代码分别如下:

Set.java

```
public interface Set {
    public void add(People people);
    public People get(int index);
```

```java
    public Iterator iterator();
    public int size();
}
```

People.java

```java
public class People {
    public String name;
    public People nextPeople;
    public People(){
        name = "无名氏";
    }
    public People(String name){
        this.name = name;
    }
    public void setNextPeople(People people){
        nextPeople = people;
    }
    public People getNextPeople(){
        return nextPeople;
    }
}
```

2．具体集合

具体集合是 Building 类。代码如下：

Building.java

```java
public class Building implements Set {
    int size;                     //集合的大小,即含有的 People 对象的个数
    People currentPeople = null;
    People head = null;           //存放集合中的第一个人
    public void add(People people){
        if(currentPeople == null){
            currentPeople = people;
            head = people;
        }
        else {
            currentPeople.setNextPeople(people);
            currentPeople = people;
        }
        size++;
    }
    public Iterator iterator() {
        return new BuidlingIterator(head);
    }
    public People get(int index) {
        if(index >= size || index < 0) {
            return null;
        }
        if(index == 0) {
            return head;
```

```
        }
        else {
            People backPeople = head;
            for(int j = 0;j < index;j++){
                backPeople = backPeople.getNextPeople();
            }
            return backPeople;
        }
    }
    public int size(){
        return size;
    }
}
```

3. 迭代器

为了学习迭代的基本原理,这里的迭代器是名字为 Iterator 的接口,但和 Java 类库中的 Iterator 接口没有关系(这里的更为简化)。Iterator 接口代码如下:

Iterator.java

```
public interface Iterator {
    public boolean hasNext();
    public People next();
}
```

4. 具体迭代器

具体迭代器是 BuildingIterator 类。代码如下:

BuildingIterator.java

```
public class BuildingIterator implements Iterator {
    People currentPeople;
    People backPeople;
    BuildingIterator(People currentPeople){
        this.currentPeople = currentPeople;
    }
    public People next(){
        if(!hasNext()) {
            return null;
        }
        else {
            backPeople = currentPeople;
            currentPeople = currentPeople.getNextPeople();
            return backPeople;
        }
    }
    public boolean hasNext(){
        return currentPeople!= null;
    }
}
```

▶ 22.2.3 模式的使用

前面已经使用迭代器模式给出了可以使用的类,这些类就是一个小框架,可以使用这个小框架中的类编写应用程序。

下列应用程序(Application.java)让集合(模拟楼房)返回迭代器(模拟警察),然后迭代器遍历集合中的全部元素(元素模拟楼房中的人)。这个应用程序也单独得到了集合中的某个元素。程序运行效果如图22.3所示。

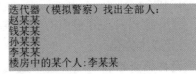

图 22.3 程序运行效果

Application.java

```java
public class Application{
    public static void main(String args[]){
        Set building = new Building();
        People zhao = new People("赵某某");
        People qian = new People("钱某某");
        People sun  = new People("孙某某");
        People li   = new People("李某某");
        building.add(zhao);
        building.add(qian);
        building.add(sun);
        building.add(li);
        Iterator police = building.iterator();        //迭代器(模拟警察)
        System.out.println("迭代器(模拟警察)找出全部人:");
        while(police.hasNext()) {
            System.out.println(police.next().name);
        }
        System.out.print("楼房中的某个人:");
        System.out.println( building.get(3).name);
    }
}
```

22.3 迭代器模式的优点

使用迭代器访问集合中的对象,不需要知道这些对象在集合中是如何表示及存储的。

22.4 应用举例——使用多个集合存储对象

扫一扫

视频讲解

1. 设计要求

Java类库提供的集合框架实现了常用的数据结构,例如链表(LinkedList)、散列映射(HashMap)、树集(TreeSet)等,并提供了相应的迭代器。链表适合插入、删除等操作;散列映射适合查找操作;树集适合排序操作。

现在有若干个学生,他们有姓名、学号和出生日期等属性,请编程完成下列四项任务:

(1) 使用链表存放学生对象,每当链表中的数据发生变化时,更新散列映射和树集中的数据。

(2) 用一个散列映射和一个树集存放链表中的对象。

（3）使用散列映射查询某个学生。

（4）通过树集将学生按成绩排序。

2. 设计实现

1）集合

集合是java.util包中的List接口、Set接口和Map接口。

2）具体集合

具体集合是java.util包中的LinkedList类、TreeSet类和HashMap类。LinkedList类实现了List接口；TreeSet类实现了Set接口；HashMap类实现了Map接口。

3）迭代器

迭代器是java.util包中的Iterator接口。

4）具体迭代器

具体迭代器是java.util包中的某些类，这些类客户无法直接使用。

5）应用程序

这里设计了三个类：UseSet、Student和Application。UseSet类包含链表、散列表和树集。Application类负责创建Student对象，并将所创建的Student对象添加到UseSet所包含的集合中。UseSet提供了按Student对象的number属性查找Student对象的方法，也提供了按Student对象的score属性进行排序的方法。程序运行效果如图22.4所示。

```
查找学号为003的学生：
学号:003  姓名:刘五  分数:58
将学生按成绩排列：
学号:003  姓名:刘五  分数:58
学号:004  姓名:赵六  分数:66
学号:001  姓名:张三  分数:76
学号:002  姓名:李四  分数:88
学号:005  姓名:周七  分数:92
删除一个学号是003的学生，添加一个学号是006的学生
再将学生按成绩排列：
学号:004  姓名:赵六  分数:66
学号:001  姓名:张三  分数:76
学号:002  姓名:李四  分数:88
学号:005  姓名:周七  分数:92
学号:006  姓名:钱六  分数:92
```

图22.4　程序运行效果

Application.java

```java
public class Application{
    public static void main(String args[]){
        UseSet useSet = new UseSet();
        useSet.addStudent(new Student("001","张三",76));
        useSet.addStudent(new Student("002","李四",88));
        useSet.addStudent(new Student("003","刘五",58));
        useSet.addStudent(new Student("004","赵六",66));
        useSet.addStudent(new Student("005","周七",92));
        String n = "003";
        System.out.println("查找学号为"+n+"的学生:");
        useSet.lookStudent(n);
        System.out.println("将学生按成绩排列:");
        useSet.printStudentsByScore();
        String m = "006";
```

```java
            System.out.println
            ("删除一个学号是" + n + "的学生,添加一个学号是" + m + "的学生");
            useSet.delete(n);
            useSet.addStudent(new Student(m,"钱六",92));
            System.out.println("再将学生按成绩排列:");
            useSet.printStudentsByScore();
    }
}
```

Student.java

```java
import java.util.*;
public class Student implements Comparable<Student>{
    String number,name;
    int score = 0;
    private int x = 10;
    public Student(){}
    Student(String number,String name,int score){
        this.number = number;
        this.name = name;
        this.score = score;
    }
    public int compareTo(Student b){
        if(this.score - b.score != 0)
            return this.score - b.score;
        else
            return 1;                    //集合中允许出现分数相同的元素
    }
    public String getNumber(){
        return number;
    }
    public String getName(){
        return name;
    }
    public int getScore(){
        return score;
    }
}
```

UseSet.java

```java
import java.util.*;
public class UseSet {
    LinkedList<Student> list;
    HashMap<String,Student> map;
    TreeSet<Student> tree;
    public UseSet(){
        list = new LinkedList<Student>();
        map = new HashMap<String,Student>();
        tree = new TreeSet<Student>();
    }
    public void addStudent(Student stu){
```

```java
            if(!map.containsKey(stu.getNumber()))
                list.add(stu);
            update();
        }
        public void delete(String number) {
            for(Student stu:list){
                if(stu.getNumber().equals(number)){
                    list.remove(stu);
                    break;
                }
            }
            update();
        }
        public void lookStudent(String num){
            Student stu = map.get(num);
            String number = stu.getNumber();
            String name = stu.getName();
            int score = stu.getScore();
            System.out.println("学号:" + number + " 姓名:" + name + " 分数:" + score);
        }
        public void printStudentsByScore(){
            Iterator<Student> iterator = tree.iterator();          //迭代器
            while(iterator.hasNext()){
                Student stu = iterator.next();
                String number = stu.getNumber();
                String name = stu.getName();
                int score = stu.getScore();
                System.out.println("学号:" + number + " 姓名:" + name + " 分数:" + score);
            }
        }
        private void update(){
            tree.clear();
            Iterator<Student> iterator = list.iterator();
            while(iterator.hasNext()){
                Student stu = iterator.next();
                String number = stu.getNumber();
                map.put(number,stu);
                tree.add(stu);
            }
        }
    }
}
```

第 23 章　组合模式

以下文本框中的内容引自 GoF 所著 *Design Patterns：Elements of Reusable Object Oriented Software* 的中译本及英文版。

> **组合模式**
> 　　将对象组合成树状结构以表示"部分-整体"的层次结构。组合模式使用户对单个对象和组合对象的使用具有一致性。
> **Composite Pattern**
> 　　Compose objects into tree structures to represent part whole hierarchies. Composite lets clients treat individual objects and compositions of objects uniformly.

以上内容是 GoF 对组合模式的高度概括，结合 23.2.1 节的组合模式的类图可以准确地理解该模式。

23.1　概述

如果一个对象包含另一个对象，称这样的对象为组合式对象。如果将当前组合式对象作为一个整体，那么它所包含的对象就是该整体的一部分。如果一个对象不含有其他对象，称这样的对象为叶式对象。在编写程序时，我们希望将多个叶式对象和组合式对象组成树形结构，以此表示"部分-整体"的层次结构。也就是说，在树形结构中，组合式对象所含有的对象将作为该组合对象的子节点被对待，并借助该层次结构使用户能用一致的方式处理叶式对象和组合式对象。在组成的树形结构中，叶式对象和组合式对象都是树中的节点，但是，组合式对象是具有其他子节点的节点，叶式对象是不具有其他子节点的叶节点，如图 23.1 所示。

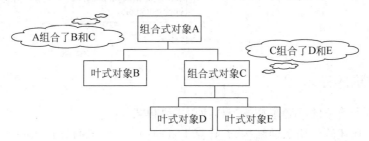

图 23.1　组成树形结构的对象

组合模式是关于怎样将对象形成树形结构来表现"部分-整体"的层次结构的成熟模式。使用组合模式，可以让用户以一致的方式处理叶式对象和组合式对象。组合模式的关键在于无论是叶式对象还是组合式对象，都实现了相同的接口，或都是同一个抽象类的子类。

23.2 模式的结构与使用

组合模式包括三种角色。

1. 抽象组件(Component)

抽象组件是一个接口(抽象类),该接口(抽象类)定义了个体对象和组合对象需要实现的关于操作其子节点的方法,例如 add()、remove()以及 getChild()等。抽象组件也可以定义个体对象和组合对象用于操作其自身的方法,例如 isLeaf()等。抽象组件也称作节点。

2. 组合式节点(Composite Node)

组合式节点是实现(扩展)抽象组件的类,其实例是组合式对象。组合式节点含有其他组合式节点或叶节点。

3. 叶节点(Leaf Node)

叶节点是实现(扩展)抽象组件的类,其实例是叶式对象。叶节点不含其他组合式节点或叶节点。

▶ 23.2.1 组合模式的 UML 类图

组合模式的类图如图 23.2 所示。

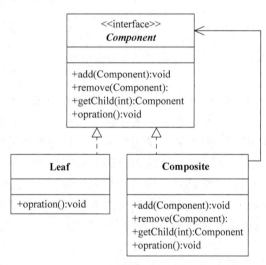

图 23.2 组合模式的类图

▶ 23.2.2 结构的描述

下面通过一个简单的问题来描述怎样使用组合模式。

用组合模式描述软件团队的结构,并计算工资(月)。一个软件团队由一个项目经理、一个设计团队(队长负责)、一个测试团队(队长负责)组成。设计团队包括三个小组,每个小组(组长负责)有若干人;测试团队包括两个小组(组长负责),每个组有若干人。使用组合模式,让软件团队的人员形成树形结构。计算某个小组的工资、设计团队的工资、测试团队的工资,以及整个软件团队的工资。

1. 抽象组件

抽象组件是 Person 接口。代码如下:

Person.java

```java
import java.util.Iterator;
public interface Person {
    public void add(Person person) ;
    public void remove(Person person) ;
    public Person getChild(int index);
    public Iterator<Person> getAllChildren() ;
    public boolean isLeaf();
    public double getSalary();
    public void setSalary(double salary);
}
```

2. 组合式节点

组合式节点是 Officer 类。代码如下:

Officer.java

```java
import java.util.LinkedList;
import java.util.Iterator;
public class Officer implements Person{
    LinkedList<Person> list;          //用 list 存放子节点
    String name;
    double salary;
    Officer(String name,double salary){
        this.name = name;
        this.salary = salary;
        list = new LinkedList<Person>();
    }
    public void add(Person person) {
        list.add(person);             //添加子节点
    }
    public void remove(Person person){
        list.remove(person);
    }
    public Person getChild(int index) {
        return list.get(index);
    }
    public Iterator<Person> getAllChildren() {
        return list.iterator();
    }
    public boolean isLeaf(){
        return false;
    }
    public double getSalary(){
        return salary;
    }
    public void setSalary(double salary){
        this.salary = salary;
    }
}
```

3. 叶节点

叶节点是 Worker 类。代码如下：

Worker.java

```java
import java.util.Iterator;
public class Worker implements Person{
    double salary;
    String name;
    Worker(String name,double salary){
        this.name = name;
        this.salary = salary;
    }
    public void add(Person person){
        System.out.println("是叶节点,不能添加子节点");
    }
    public void remove (Person person){
        System.out.println("是叶节点,没有子节点可删除");
    }
    public Person getChild(int index) {
        return null;
    }
    public Iterator< Person > getAllChildren() {
        return null;
    }
    public boolean isLeaf(){
        return true;
    }
    public double getSalary(){
        return salary;
    }
    public void setSalary(double salary){
        this.salary = salary;
    }
}
```

23.2.3 模式的使用

前面已经使用组合模式给出了可以使用的类,这些类就是一个小框架,可以使用这个小框架中的类编写应用程序。

下列应用程序(Application.java)使用了组合模式中所涉及的类,通过确定"部分-整体"关系来计算工资。程序运行效果如图 23.3 所示。

```
设计团队的工资:606000.0
测试团队的工资:440000.0
设计团队一组的工资:195600.0
测试团队二组的工资:209400.0
整个软件团队的工资:1065000.0
```

图 23.3 程序运行效果

Application.java

```java
public class Application{
    public static void main(String args[]) {
        Person
        manager = new Officer("项目经理",19000),        //组合式节点
```

第23章 组合模式

```java
            designTeamLeader = new Officer("设计团队队长",18000);
            testTeamLeader = new Officer("测试团队队长",13000);
            designTeamGroupLeader1 = new Officer("设计团队组长 1",15600);
            designTeamGroupLeader2 = new Officer("设计团队组长 2",16600);
            designTeamGroupLeader3 = new Officer("设计团队组长 3",15800);
            testTeamGroupLeader1 = new Officer("测试团队组长 1",12900);
            testTeamGroupLeader2 = new Officer("测试团队组长 2",13600);
            worker [] = new Worker[90];
            for(int i = 0;i < worker.length;i++){
                if(i % 2 == 0)
                    worker[i] = new Worker("设计人员",12000);        //叶节点
                else
                    worker[i] = new Worker("测试人员",8900);         //叶节点
            }
            manager.add(designTeamLeader);                           //经理组合设计团队队长
            manager.add(testTeamLeader);                             //经理组合测试团队队长
            designTeamLeader.add(designTeamGroupLeader1);            //设计团队队长组合设计组组长 1
            designTeamLeader.add(designTeamGroupLeader2);            //设计团队队长组合设计组组长 2
            designTeamLeader.add(designTeamGroupLeader3);            //设计团队队长组合设计组组长 3
            testTeamLeader.add(testTeamGroupLeader1);                //测试团队队长组合测试组组长 1
            testTeamLeader.add(testTeamGroupLeader2);                //测试团队队长组合测试组组长 2
            for(int m = 0;2 * m < worker.length;m++){                //组长开始组合开发人员
                if(m % 3 == 0)
                    designTeamGroupLeader1.add(worker[2 * m]);
                else if(m % 3 == 1)
                    designTeamGroupLeader2.add(worker[2 * m]);
                else if(m % 3 == 2)
                    designTeamGroupLeader3.add(worker[2 * m]);
            }
            for(int m = 0;2 * m + 1 < worker.length;m++){            //组长开始组合测试人员
                if(m % 2 == 0)
                    testTeamGroupLeader1.add(worker[2 * m + 1]);
                else if(m % 2 == 1)
                    testTeamGroupLeader2.add(worker[2 * m + 1]);
            }
            System.out.println
            ("设计团队的工资:" + ComputerSalary.computerSalary(designTeamLeader));
            System.out.println
            ("测试团队的工资:" + ComputerSalary.computerSalary(testTeamLeader));
            System.out.println
            ("设计团队一组的工资:" + ComputerSalary.computerSalary(designTeamGroupLeader1));
            System.out.println
            ("测试团队二组的工资:" + ComputerSalary.computerSalary(testTeamGroupLeader2));
            System.out.println("整个软件团队的工资:" + ComputerSalary.computerSalary(manager));
    }
}
```

ComputerSalary.java

```java
import java.util.Iterator;
public class ComputerSalary{
```

```java
public static double computerSalary(Person person){
    double sum = 0;
    if(person.isLeaf() == true){
        sum = sum + person.getSalary();
    }
    if(person.isLeaf() == false){
        sum = sum + person.getSalary();
        Iterator < Person > iterator = person.getAllChildren();
        while(iterator.hasNext()){
            Person child = iterator.next();
            sum = sum + computerSalary(child);        //递归调用
        }
    }
    return sum;
}
```

23.3 组合模式的优点

组合模式中包含叶节点和组合式节点,并形成树形结构,使用户可以方便地处理叶节点和组合式节点。叶节点和组合式节点实现了相同的接口,用户一般无须区分叶节点和组合式节点。当增加新的组合式节点或叶节点时,用户的重要代码不需要做出修改。

23.4 应用举例——苹果树的苹果价值

1. 设计要求

一棵苹果树的主干上有两个分支,一个分支上结了 10 个苹果,另一个分支上结了 8 个苹果,苹果 12 元/kg。请用组合模式组织苹果树的结构,计算苹果树上苹果的重量和苹果的价值。

2. 设计实现

1) 抽象组件

抽象组件是 TreeComponent 类。代码如下:

TreeComponent.java

```java
import java.util.Iterator;
public interface TreeComponent{
    public void add(TreeComponent node);
    public void remove(TreeComponent node);
    public TreeComponent getChild(int index);
    public Iterator < TreeComponent > getAllChildren();
    public boolean isLeaf();
    public double getWeight();
}
```

2) 组合式节点

组合式节点是 TreeBody 类。代码如下:

TreeBody.java

```java
import java.util.LinkedList;
import java.util.Iterator;
public class TreeBody implements TreeComponent{
    LinkedList<TreeComponent> list;          //存放节点
    String name;
    TreeBody(String name){
        this.name = name;
        list = new LinkedList<TreeComponent>();
    }
    public void add(TreeComponent node) {
        list.add(node);
    }
    public void remove(TreeComponent node){
        list.remove(node);
    }
    public TreeComponent getChild(int index) {
        return list.get(index);
    }
    public Iterator<TreeComponent> getAllChildren() {
        return list.iterator();
    }
    public boolean isLeaf(){
        return false;
    }
    public double getWeight(){
        return 0;
    }
    public String toString(){
        return name;
    }
}
```

3）叶结点

叶节点是 Apple 类。代码如下：

Apple.java

```java
import java.util.Iterator;
public class Apple implements TreeComponent{
    double weight;
    String name;
    Apple(String name,double weight){
        this.name = name;
        this.weight = weight;
    }
    public void add(TreeComponent node) {}
    public void remove(TreeComponent node){}
    public TreeComponent getChild(int index) {
        return null;
    }
```

```
        public Iterator < TreeComponent > getAllChildren() {
            return null;
        }
        public boolean isLeaf(){
            return true;
        }
        public double getWeight(){
            return weight;
        }
        public String toString(){
            return name;
        }
}
```

4）应用程序

下列应用程序（Application.java）使用了组合模式中所涉及的类，确定"部分-整体"关系，计算苹果树上苹果的重量和价值。此外，这个应用程序还使用 Java API 提供的 JTree 组件为组合模式的树形结构提供了视图。程序运行效果如图 23.4 所示。

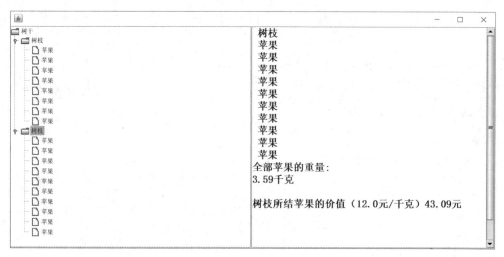

图 23.4　程序运行效果

Application.java

```
import javax.swing. * ;
import javax.swing.tree. * ;
import javax.swing.event. * ;
import java.awt. * ;
public class Application extends JFrame implements TreeSelectionListener{
    TreeComponent mainBody,branchOne,branchTwo,apple[];            //节点
    DefaultMutableTreeNode trunk,branch1,branch2,leaf[];           //为节点提供的外观
    JTree tree;
    final static int MAX = 18;                                      //共有 18 个苹果
    JTextArea text;
    public Application() {
        mainBody = new TreeBody("树干");
```

```java
            trunk = new DefaultMutableTreeNode(mainBody);         //为 mainBody 提供的外观
            branchOne = new TreeBody("树枝");
            branch1 = new DefaultMutableTreeNode(branchOne);       //为 branchOne 提供的外观
            branchTwo = new TreeBody("树枝");
            branch2 = new DefaultMutableTreeNode(branchTwo);
            apple = new Apple[MAX];
            leaf = new DefaultMutableTreeNode[MAX];                //为 apple 提供的外观
            for(int i = 0;i < MAX;i++){
                double weight = (int)(Math.random() * 100 + 0.5)/1000.0;
                apple[i] = new Apple("苹果",weight + 0.3);
                leaf[i] = new DefaultMutableTreeNode(apple[i]);
            }
            mainBody.add(branchOne);
            trunk.add(branch1);
            mainBody.add(branchTwo);
            trunk.add(branch2);
            for(int i = 0;i <= 7;i++){
                branchOne.add(apple[i]);
                branch1.add(leaf[i]);
            }
            for(int i = 8;i < MAX;i++){
                branchTwo.add(apple[i]);
                branch2.add(leaf[i]);
            }
            tree = new JTree(trunk);
            tree.addTreeSelectionListener(this);
            text = new JTextArea(20,20);
            text.setFont(new Font("宋体",Font.BOLD,20));
            text.setLineWrap(true);
            setLayout(new GridLayout(1,2));
            add(new JScrollPane(tree));
            add(new JScrollPane(text));
            setBounds(70,80,960,520);
            setDefaultCloseOperation(JFrame.EXIT_ON_CLOSE);
            setVisible(true);
        }
        public void valueChanged(TreeSelectionEvent e){
            text.setText(null);
            DefaultMutableTreeNode node =
            (DefaultMutableTreeNode)tree.getLastSelectedPathComponent();
            TreeComponent treeComponent = (TreeComponent)node.getUserObject();
            String allName = Computer.getAllChildrenName(treeComponent);
            double weight = Computer.computerAppleWeight(treeComponent);
            String weightStr = String.format("%.2f",weight);
            String mess = null;
            if(treeComponent.isLeaf())
                mess = allName + "的重量:\n " + weightStr + "千克";
            else
                mess = allName + "\n 全部苹果的重量:\n" + weightStr + "千克";
            text.append(mess + "\n");
            double unit = 12;                                       //苹果:12 元/千克
```

```java
            double value = Computer.computerValue(treeComponent,unit);
            String valueStr = String.format("%.2f",value);
            String name = treeComponent.toString();
            if(treeComponent.isLeaf())
                mess = name + "的价值(" + unit + "元/千克)" + valueStr + "元";
            else
                mess = name + "所结苹果的价值(" + unit + "元/千克)" + valueStr + "元";
            text.append("\n" + mess);
        }
        public static void main(String args[]) {
            new Application();
        }
}
```

Computer.java

```java
import java.util.Iterator;
public class Computer{
    public static double computerAppleWeight(TreeComponent node){        //求树枝上苹果的重量
        double weightSum = 0;
        if(node.isLeaf() == true){
            weightSum = weightSum + node.getWeight();
        }
        if(node.isLeaf() == false){
            Iterator<TreeComponent> iterator = node.getAllChildren();
            while(iterator.hasNext()){
                TreeComponent child = iterator.next();
                weightSum = weightSum + computerAppleWeight(child);      //递归调用
            }
        }
        return weightSum;
    }
    public static double computerValue(TreeComponent node,double unit){
        //求树枝上苹果的价值
        double appleWorth = 0;
        if(node.isLeaf() == true){
            appleWorth = appleWorth + node.getWeight() * unit;
        }
        if(node.isLeaf() == false){
            Iterator<TreeComponent> iterator = node.getAllChildren();
            while(iterator.hasNext()){
                TreeComponent child = iterator.next();
                appleWorth = appleWorth + computerValue(child,unit);     //递归调用
            }
        }
        return appleWorth;
    }
    public static String getAllChildrenName(TreeComponent node){
        StringBuffer mess = new StringBuffer();
        if(node.isLeaf() == true){
            mess.append(" " + node.toString());
```

```java
        }
        if(node.isLeaf() == false){
            mess.append(" " + node.toString());
            Iterator < TreeComponent > iterator = node.getAllChildren();
            while(iterator.hasNext()){
                TreeComponent child = iterator.next();
                mess.append("\n" + getAllChildrenName(child));         //递归调用
            }
        }
        return new String(mess);
    }
}
```

第 24 章 观察者模式

以下文本框中的内容引自 GoF 所著 *Design Patterns：Elements of Reusable Object Oriented Software* 的中译本及英文版。

> **观察者模式（别名：依赖，发布，订阅）**
> 定义对象间的一种一对多的依赖关系，当一个对象的状态发生变化时，所有依赖它的对象都得到通知并被自动更新。
> **Observer Pattern（Another Name：Dependents，Publish，Subscribe）**
> Define a one to many dependency between objects so that when one object changes state, all its dependents are notified and updated automatically.

以上内容是 GoF 对观察者模式的高度概括，结合 24.2.1 节的观察者模式的类图可以准确地理解该模式。

24.1 概述

在许多设计中，经常涉及多个对象都对一个特殊对象的数据变化感兴趣，而且这些对象都希望跟踪那个特殊对象的数据变化。例如，某些求职者对"求职中心"的职位需求信息的变化非常关心，很想跟踪"求职中心"中职位需求信息的变化。他们要首先成为"求职中心"的"求职者"，即让"求职中心"把自己登记到"求职中心"的"求职者"列表中，然后"求职中心"就会及时通知他最新的职位需求信息。如果一个"求职者"不想继续知道"求职中心"的职位需求信息，就让"求职中心"把自己从"求职中心"的"求职者"列表中删除，"求职中心"就不会再通知他职位需求信息。"求职中心"和"求职者"的关系示意如图 24.1 所示。

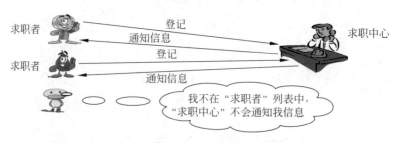

图 24.1 "求职中心"与"求职者"的关系

观察者模式是关于多个对象想知道一个对象中数据变化情况的一种成熟的模式。观察者模式中有一个称作"主题"的对象和若干个称作"观察者"的对象。"主题"和"观察者"之间是一种一对多的依赖关系，当"主题"的状态发生变化时，所有"观察者"都得到通知。前面所述的"求职中心"相当于观察者模式的一个具体"主题"；每个"求职者"相当于观察者模式中的一个具体"观察者"。

24.2 模式的结构与使用

观察者模式的结构中包括四种角色。

1. 主题（Subject）

主题是一个接口，该接口规定了具体主题需要实现的方法，例如添加观察者、删除观察者以及通知观察者更新数据。

2. 观察者（Observer）

观察者是一个接口，该接口规定了具体观察者用来更新数据的方法。

3. 具体主题（Concrete Subject）

具体主题是实现主题接口的类。具体主题需要使用一个集合，例如 ArrayList，存放具体观察者，以便数据变化时通知具体观察者。

4. 具体观察者（Concrete Observer）

具体观察者是实现观察者接口的类。具体观察者组合主题，以便让自己成为主题的观察者，或让自己不再是主题的观察者。

▶ 24.2.1 观察者模式的 UML 类图

观察者模式的类图如图 24.2 所示。

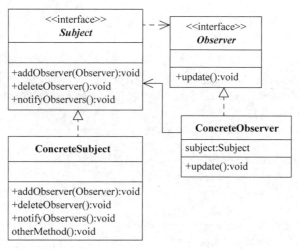

图 24.2 观察者模式的类图

▶ 24.2.2 结构的描述

下面通过一个简单的问题来描述观察者模式中所涉及的各个角色。

一个大学生和一个归国留学生都希望及时知道"求职中心"最新的职位需求信息。

1. 主题

主题是 Subject 接口，该接口规定了具体主题需要实现的添加观察者、删除观察者以及发送数据给观察者的方法。代码如下：

Subject.java

```java
public interface Subject{
    public void addObserver(Observer o);
    public void deleteObserver(Observer o);
    public void notifyObservers(String str);
}
```

2. 观察者

观察者是 Observer 接口。该接口规定了具体观察者用来更新数据的方法。对于本问题，观察者接口规定的方法是 hearTelephone()（相当于观察者模式类图中的 update()方法），即要求具体观察者通过实现 hearTelephone()方法（模拟接听电话）来更新数据。Observer 接口的代码如下：

Observer.java

```java
public interface Observer{
    public void hearTelephone(String heardMess);
}
```

3. 具体主题

具体主题是 SeekJobCenter 类。具体主题通过实现 notifyObservers()方法来通知具体观察者，实现的方式是遍历具体主题中用来存放观察者的集合，并让集合中的每个具体观察者执行观察者接口（Observer）规定更新数据的方法，例如 hearTelephone()方法。SeekJobCenter 类的代码如下：

SeekJobCenter.java

```java
import java.util.ArrayList;
public class SeekJobCenter implements Subject{
    ArrayList<Observer> personList;              //存放观察者对象的数组表
    public SeekJobCenter(){
        personList = new ArrayList<Observer>();
    }
    public void addObserver(Observer observer){
        if(!(personList.contains(observer)))
            personList.add(observer);            //把观察者对象添加到数组表
    }
    public void deleteObserver(Observer observer){
        if(personList.contains(observer))
            personList.remove(observer);
    }
    public void notifyObservers(String mess){    //通知所有的观察者
        for(int i = 0;i < personList.size();i++){
            Observer observer = personList.get(i);
            observer.hearTelephone(mess);        //让观察者接听电话
        }
    }
}
```

4. 具体观察者

具体观察者是 Student 类和 HaiGui 类。Student 类的实例调用 hearTelephone(String heardMess)方法时,会将收到的数据保存到一个文件中。HaiGui 类的实例调用 hearTelephone(String heardMess)方法时,如果数据中包含"程序员"或"软件"等字样,就将信息保存到一个文件中。Student 类和 HaiGui 类的代码分别如下:

Student. java

```java
import java.io.*;
public class Student implements Observer{
    Subject subject;
    File myFile;
    public Student(Subject subject,String fileName){
        this.subject = subject;
        subject.addObserver(this);          //使当前对象成为 subject 具体主题的观察者
        myFile = new File(fileName);
    }
    public void hearTelephone(String heardMess){
        try{
            RandomAccessFile out = new RandomAccessFile(myFile,"rw");
            out.seek(out.length());
            byte b[] = heardMess.getBytes();
            out.write(b);                   //更新文件中的内容
            System.out.print("我是一个大学生,");
            System.out.println("我向文件" + myFile.getName() + "写入如下内容:");
            System.out.println(heardMess);
        }
        catch(IOException exp){
            System.out.println(exp.toString());
        }
    }
}
```

HaiGui. java

```java
import java.io.*;
public class HaiGui implements Observer{
    Subject subject;
    File myFile;
    public HaiGui(Subject subject,String fileName){
        this.subject = subject;
        subject.addObserver(this);          //使当前对象成为 subject 具体主题的观察者
        myFile = new File(fileName);
    }
    public void hearTelephone(String heardMess){
        try{
            boolean isOK = heardMess.contains("Java 程序员")||
                           heardMess.contains("软件");
            RandomAccessFile out = new RandomAccessFile(myFile,"rw");
            if(isOK){
                out.seek(out.length());
                byte b[] = heardMess.getBytes();
                out.write(b);               //更新文件中的内容
                System.out.print("我是一个海归,");
```

```java
                    System.out.println("我向文件" + myFile.getName() + "写入如下内容:");
                    System.out.println(heardMess);
                }
            }
            catch(IOException exp){
                System.out.println(exp.toString());
            }
        }
    }
```

▶ 24.2.3　模式的使用

前面已经使用观察者模式给出了可以使用的类,这些类就是一个小框架,可以使用这个小框架中的类编写应用程序。

下列应用程序(Application.java)使用了观察者模式中所涉及的类,演示一个大学生和一个归国留学生成为求职中心的观察者;当求职中心有新的人才需求信息时,大学生和归国留学生将得到通知。程序运行效果如图 24.3 所示。

```
我是一个大学生,我向文件A.txt写入如下内容:
星星公司需要20个Java程序员。
我是一个海归,我向文件B.txt写入如下内容:
星星公司需要20个Java程序员。
我是一个大学生,我向文件A.txt写入如下内容:
月月公司需要8个动画设计师。
我是一个大学生,我向文件A.txt写入如下内容:
星月公司需要9个电工。
```

图 24.3　程序运行效果

Application.java

```java
public class Application{
    public static void main(String args[]){
        Subject center = new SeekJobCenter();                              //具体主题
        Observer zhang = new Student(center,"A.txt");                      //具体观察者
        Observer wang = new HaiGui(center,"B.txt");                        //具体观察者
        center.notifyObservers("星星公司需要 20 个 Java 程序员.");          //主题通知信息
        center.notifyObservers("月月公司需要 8 个动画设计师.");
        center.notifyObservers("星月公司需要 9 个电工.");
    }
}
```

24.3　观察者模式的优点

具体主题和具体观察者是松耦合关系。由于主题(Subject)接口仅仅依赖于观察者(Observer)接口,因此具体主题只是知道它的观察者是实现观察者(Observer)接口的某个类的实例,但不需要知道具体是哪个类。同样,由于观察者仅仅依赖于主题(Subject)接口,因此具体观察者只是知道它依赖的主题是实现主题(Subject)接口的某个类的实例,但不需要知道具体是哪个类。

观察者模式满足"开-闭"原则。主题(Subject)接口仅仅依赖于观察者(Observer)接口,这样,就可以让创建具体主题的类也仅仅依赖于观察者(Observer)接口,因此,如果增加新的实现观察者(Observer)接口的类,不必修改创建具体主题的类的代码。同样,创建具体观察者的类仅仅依赖于主题(Observer)接口,如果增加新的实现主题(Subject)接口的类,也不必修改创建具体观察者类的代码。

24.4 应用举例——求面积服务中心

1. 设计要求

用观察者模式建立一个"求面积服务中心",该中心可以帮助用户计算面积,例如三角形的面积、圆的面积等。"求面积服务中心"属于主题角色,观察者角色帮助"求面积服务中心"完成求面积的任务。当"求面积服务中心"向所有观察者发出求某个图形的面积的请求时,其中一个观察者会完成求面积,并将结果返回给"求面积服务中心"。

2. 设计实现

1) 主题

主题是 CalculatedCenter 类。代码如下:

CalculatedCenter.java

```java
public interface CalculatedCenter{
    public void addObserver(Observer o);
    public void deleteObserver(Observer o);
    public void notifyObservers(String str);
    public void getBackMess(String str);
    public void showResult();
}
```

2) 观察者

观察者是 Observer 接口。代码如下:

Observer.java

```java
public interface Observer{
    public void receiveInformation(String data);         //接收数据
}
```

3) 具体主题

具体主题是 CalculatedAreaCenter 类。具体主题通过实现 notifyObservers()方法来通知具体观察者,实现的方式是遍历具体主题中用来存放观察者的集合,并让集合中的每个具体观察者执行观察者接口(Observer)规定更新数据的方法,例如 receiveInformation()方法。CalculatedAreaCenter 类的代码如下:

CalculatedAreaCenter.java

```java
import java.util.ArrayList;
public class CalculatedAreaCenter implements CalculatedCenter{
    String result;
    ArrayList<Observer> personList;          //存放观察者对象的数组表
    public CalculatedAreaCenter(){
        personList = new ArrayList<Observer>();
    }
    public void addObserver(Observer observer){
        if(!(personList.contains(observer))) {
            personList.add(observer);        //把观察者对象添加到数组表
```

```java
        }
    }
    public void deleteObserver(Observer observer){
        if(personList.contains(observer)) {
            personList.remove(observer);
        }
    }
    public void notifyObservers(String mess){           //通知所有的观察者
        for(int i = 0;i < personList.size();i++){
            Observer observer = personList.get(i);
            observer.receiveInformation(mess);          //让观察者接受信息
        }
    }
    public void getBackMess(String str){
        result = str;
    }
    public void showResult(){
        System.out.println(result);
    }
}
```

4）具体观察者

具体观察者是 TriangleObserver 类和 CircleObserver 类。TriangleObserver 类的实例调用 receiveInformation（String data）方法时，会根据收到的数据，计算三角形的面积，并将面积返回给主题。CircleObserver 类的实例调用 receiveInformation（String data）方法时，会根据收到的数据，计算圆的面积，并将面积返回给主题。TriangleObserver 类和 CircleObserver 类的代码分别如下：

TriangleObserver.java

```java
import java.util.regex.Pattern;
import java.util.regex.Matcher;
public class TriangleObserver implements Observer{
    Pattern pattern;                                    //模式对象
    Matcher matcher;                                    //匹配对象
    double sideA ,sideB,sideC;
    CalculatedCenter subject;
    public TriangleObserver(CalculatedCenter su){
        subject = su;
        subject.addObserver(this);                      //使当前对象成为 subject 具体主题的观察者
    }
    public void receiveInformation(String data){        //接收数据
        if(data.contains("三角形")){
            String regex = "-?[0-9][0-9]*[.]*[0-9]*";   //匹配浮点数,包括整数
            pattern = Pattern.compile(regex);           //初始化模式对象
            matcher = pattern.matcher(data);
            if(matcher.find())
                sideA = Double.parseDouble(matcher.group());
            if(matcher.find())
                sideB = Double.parseDouble(matcher.group());
```

```java
            if(matcher.find())
                sideC = Double.parseDouble(matcher.group());
            double p = (sideA + sideB + sideC)/2.0;
            double area = Math.sqrt(p * (p - sideA) * (p - sideB) * (p - sideC));
            subject.getBackMess
            ("三边为" + sideA + "," + sideB + "," + sideC + "的三角形的面积是:" + area);
                                                                                    //返回结果
        }
    }
}
```

CircleObserver.java

```java
import java.util.regex.Pattern;
import java.util.regex.Matcher;
public class CircleObserver implements Observer{
    Pattern pattern;                                    //模式对象
    Matcher matcher;                                    //匹配对象
    double radius;
    CalculatedCenter subject;
    public CircleObserver(CalculatedCenter su){
        subject = su;
        subject.addObserver(this);                      //使当前对象成为subject具体主题的观察者
    }
    public void receiveInformation(String data){        //接收数据
        if(data.contains("圆")){
            String regex = "-?[0-9][0-9]*[.]*[0-9]*";   //匹配浮点数,包括整数
            pattern = Pattern.compile(regex);           //初始化模式对象
            matcher = pattern.matcher(data);
            if(matcher.find())
                radius = Double.parseDouble(matcher.group());
            double area = Math.PI * radius * radius ;
            subject.getBackMess
            ("半径为" + radius + "的圆的面积是:" + area);
                                                        //返回结果
        }
    }
}
```

5) 应用程序

下列应用程序(Application.java)使用观察者模式给出的类,计算三角形和圆的面积。程序运行效果如图24.4所示。

```
三边为4.0,5.0,6.0的三角形的面积是: 9.921567416492215
半径为67.9的圆的面积是: 14484.030186036922
```

图 24.4 程序运行效果

Application.java

```java
public class Application{
    public static void main(String args[]){
        CalculatedCenter center = new CalculatedAreaCenter();           //具体主题
```

```
        Observer triangle = new TriangleObserver(center);
        Observer circle = new CircleObserver(center);
        center.notifyObservers("计算三边是 4.0,5.0,6.0 的三角形的面积.");      //通知信息
        center.showResult();
        center.notifyObservers("计算半径是 67.9 的圆的面积。");              //通知信息
        center.showResult();
    }
}
```

第 25 章　原型模式

以下文本框中的内容引自 GoF 所著 *Design Patterns*：*Elements of Reusable Object Oriented Software* 的中译本及英文版。

> **原型模式**
> 　　用原型实例指定创建对象的种类，并且通过复制这些原型创建新的对象。
>
> **Prototype Pattern**
> 　　Specify the kinds of objects to create using a prototypical instance, and create new objects by copying this prototype.

以上内容是 GoF 对原型模式的高度概括，结合 25.2.1 节的原型模式的类图可以准确地理解该模式。

25.1　概述

在某些情况下，不希望反复使用类的构造方法创建许多对象，而是希望用该类创建一个对象后，以该对象为原型得到该对象的若干个复制品。也就是说，将一个对象定义为原型对象，要求该原型对象提供一个方法，该原型对象调用此方法可以复制一个和自己有完全相同状态的同类型的对象，即该方法"克隆"原型对象得到一个新对象，称作原型对象的复制品。这里使用"克隆"一词比"复制"更为形象。原型对象与以它为原型"克隆"出的复制品可以分别独立地变化，也就是说，改变原型对象的状态不会影响到复制品，改变复制品的状态也不会影响到原型对象。例如，通过复制一个已有的 Word 文档中的文本创建一个新的 Word 文档后，两个文档中的文本内容可以独立地变化、互不影响，也就是说，改变一个含有文本数据的原型对象含有的文本数据不会影响以它为原型克隆出的复制品，如图 25.1 所示。

图 25.1　原型和它的两个复制品

原型模式是从一个对象出发得到一个和自己有相同状态的新对象的成熟模式，该模式的关键是将一个对象定义为原型，并为其提供复制自己的方法。

25.2 模式的结构与使用

原型模式的结构中包括两种角色。

1. 抽象原型（Prototype）

抽象原型是一个接口，负责定义对象复制自身的方法。

2. 具体原型（Concrete Prototype）

具体原型是实现抽象原型的类。具体原型实现抽象原型中的方法，以便所创建的对象调用该方法复制自己。

25.2.1 原型模式的 UML 类图

原型模式的类图如图 25.2 所示。

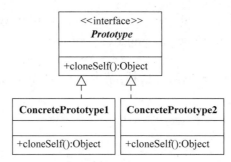

图 25.2 原型模式的类图

Java 实现克隆有两种办法。

1. 使用 java.lang 包的 Cloneable 接口

Cloneable 接口中没有任何方法，该接口的唯一作用是让 JVM 知道实现该接口的对象是可以被克隆的。一个类实现 Cloneable 接口，然后重写 Object 类的 protected Object clone() 方法，而且要首先调用覆盖掉的 clone() 方法（super.clone()），那么该类的对象就可以使用这个 clone() 方法复制自己。其代码如下所示：

```
public class A implements Cloneable {                    //实现 Cloneable 接口
    protected A clone() throws CloneNotSupportedException{   //重写 clone 方法
        A object = (A) super.clone();
        return object;
    }
}
```

注意：如果对象是组合式对象，被组合的对象也都要实现 Cloneable 接口，并重写 Object 类的 clone() 方法。使用 Cloneable 接口经常涉及深度克隆问题，使用不当会导致克隆失败。如果不是组合式对象，建议使用 Cloneable 接口。

2. 使用 java.io 包中的 Serializable 接口

相对于 clone() 方法，Java 又提供了另一种解决方案，即使用 Serializable 接口和对象流克隆对象。一个类如果实现了 Serializable 接口（java.io 包中的接口），那么这个类创建的对象就是所谓序列化的对象。需要强调的是，Serializable 接口中没有任何方法，该接口的唯一作用是让 JVM 知道实现该接口的对象是可以被序列化的。因此,实现该接口的类不需要去编写实现 Serializable 接口中的方法的代码。

使用对象输出流将该对象写入目的地,然后再将该目的地作为一个对象输入流的源,那么该对象输入流从源中读回的对象一定是原对象的一个复制品,即对象输入流通过对象的序列化信息来得到当前对象的一个克隆。

注意:使用对象流把一个对象写入对象输出流时不仅要保证该对象是序列化的,而且该对象的成员对象也必须是序列化的。如果是组合式对象,建议使用 Serializable 接口。

▶ 25.2.2 结构的描述

下面通过一个简单的问题来描述原型模式中所涉及的各个角色。

通过克隆一个矩形和一个山羊来描述原型模式中所涉及的各个角色。

1. 抽象原型

抽象原型是 PrototypeShape 接口和 PrototypeAnimal 接口。代码分别如下:

PrototypeShape.java

```java
public interface PrototypeShape {
    public PrototypeShape cloneMe();
    public int getW();
    public int getH();
    public void setW(int w);
    public void setH(int h);
}
```

PrototypeAnimal.java

```java
public interface PrototypeAnimal {
    public PrototypeAnimal cloneMe();
    public StringBuffer getName();
}
```

2. 具体原型

具体原型是 Rectangle 类和 Goat 类。Rectangle 类额外实现 Cloneable 接口;Goat 类额外实现 Serializable 接口。注意:StringBuffer 类是 java.lang 包中的实现了 Serializable 接口的 final 类。二者的代码分别如下:

Rectangle.java

```java
//CloneNotSupportedException 是 java.lang 包中的类
public class Rectangle implements Cloneable,PrototypeShape {
    int width = 160,
        height = 120;                          //矩形的宽
                                               //矩形的高
    protected PrototypeShape clone() throws CloneNotSupportedException{
        PrototypeShape rectCopy = (PrototypeShape)super.clone();
        return rectCopy;
    }
    public PrototypeShape cloneMe(){
```

```java
        PrototypeShape rectCopy = null;
        try {
            rectCopy = clone();
        }
        catch(CloneNotSupportedException exp){}
        return rectCopy;
    }
    public int getW(){
        return width;
    }
    public int getH(){
        return height;
    }
    public void setW(int w){
        width = w;
    }
    public void setH(int h){
        height = h;
    }
}
```

Goat.java

```java
//StringBuffer 是 java.lang 包中的类,实现了 Serializable 接口
import java.io.*;
public class Goat implements PrototypeAnimal,Serializable {
    public StringBuffer name = new StringBuffer("白色山羊");
    public PrototypeAnimal cloneMe() {
        PrototypeAnimal goatCopy = null;
        try{
            ByteArrayOutputStream outOne = new ByteArrayOutputStream();
            ObjectOutputStream outTwo = new ObjectOutputStream(outOne);
            outTwo.writeObject(this);                              //将原型对象写入对象输出流
            ByteArrayInputStream inOne =
            new ByteArrayInputStream(outOne.toByteArray());
            ObjectInputStream inTwo = new ObjectInputStream(inOne);
            goatCopy = (PrototypeAnimal)inTwo.readObject();   // goatCopy 就是原型的复制品
        }
        catch(Exception event){
        }
        return goatCopy;
    }
    public StringBuffer getName() {
        return name;
    }
}
```

▶ **25.2.3 模式的使用**

前面已经使用原型模式给出了可以使用的类,这些类就是一个小框架,可以使用这个小框

架中的类编写应用程序。

在 Application.java 应用程序中,首先使用 Rectangle 类和 Goat 类分别创建一个原型对象,然后原型对象各自使用接口的 cloneMe() 方法得到自己的复制品对象。程序运行效果如图 25.3 所示。

```
原型矩形的长160,宽120。
复制品长160,宽120。
将复制品长和宽更改为200,300。
原型矩形的长和宽仍然是160,120。
原型山羊的名字是白色山羊。
复制品的名字是白色山羊。
原型山羊将自己的名字改成黑色山羊。
复制品的名字还是白色山羊。
```

图 25.3　程序运行效果

Application.java

```java
public class Application{
    public static void main(String args[]){
        PrototypeShape rect = new Rectangle(),      //矩形原型
                       rectCopy = null;              //原型的复制品
        System.out.printf("原型矩形的长%d,宽%d。\n",rect.getW(),rect.getH());
        rectCopy = rect.cloneMe();                   //得到原型的复制品
        System.out.printf
              ("复制品长%d,宽%d。\n",rectCopy.getW(),rectCopy.getH());
        rectCopy.setW(200);
        rectCopy.setH(300);
        System.out.printf
              ("将复制品长和宽更改为%d,%d。\n",rectCopy.getW(),rectCopy.getH());
        System.out.printf
              ("原型矩形的长和宽仍然是%d,%d。\n",rect.getW(),rect.getH());
        PrototypeAnimal goat = new Goat();           //山羊原型
        System.out.println("原型山羊的名字是" + goat.getName() + "。");
        PrototypeAnimal goatCopy = goat.cloneMe();   //得到原型的复制品
        System.out.println("复制品的名字是" + goatCopy.getName() + "。");
        goat.getName().setCharAt(0,'黑');
        System.out.println("原型山羊将自己的名字改成" + goat.getName() + "。");
        System.out.println("复制品的名字还是" + goatCopy.getName() + "。");
    }
}
```

25.3　原型模式的优点

原型模式具有以下优点:

(1) 当创建类的新实例的代价更大时,使用原型模式复制一个已有的实例可以提高创建新实例的效率。

(2) 可以动态地保存当前对象的状态。在运行时,可以随时使用对象流保存当前对象的一个复制品。

(3) 可以在运行时创建新的对象,而无须创建一系列类和继承结构。

(4) 可以动态地添加、删除原型的复制品。

25.4　应用举例——克隆容器

1. 设计要求

在一个窗口中有一个容器,该容器中有若干个按钮组件,用户单击按钮可为该按钮选择一个背景颜色。当用户为所有按钮选定颜色后,希望复制当前容器,并把这个复制品也添加到当前窗口中(Java 组件相关的类,例如 Component 等,都已经实现了 Serializable 接口)。

视频讲解

2．设计实现

1）抽象原型

抽象原型是 CloneContainer 接口。代码如下：

CloneContainer.java

```java
public interface CloneContainer {
    public CloneContainer cloneContainer();
}
```

2）具体原型

具体原型是 ButtonContainer 类，ButtonContainer 类使用对象序列化来复制对象。代码如下：

ButtonContainer.java

```java
import java.io.*;
import javax.swing.*;
import java.awt.*;
public class ButtonContainer extends JPanel implements CloneContainer {
    JButton button[];
    public ButtonContainer(){
        button = new JButton[25];
        setLayout(new GridLayout(5,5));
        for(int i = 0;i < 25;i++){
            button[i] = new JButton();
            add(button[i]);
            button[i].addActionListener((e) ->{
                JButton b = (JButton)e.getSource();
                Color newColor =
                JColorChooser.showDialog(null,"",b.getBackground());
                if(newColor != null)
                    b.setBackground(newColor);
            });
        }
    }
    public CloneContainer cloneContainer() {            //实现原型接口中的方法
        CloneContainer objectCopy = null;               //复制品
        try{
            ByteArrayOutputStream outOne = new ByteArrayOutputStream();
            ObjectOutputStream outTwo = new ObjectOutputStream(outOne);
            outTwo.writeObject(this);                   //将原型对象写入对象输出流
            ByteArrayInputStream inOne =
            new ByteArrayInputStream(outOne.toByteArray());
            ObjectInputStream inTwo = new ObjectInputStream(inOne);
            objectCopy = (CloneContainer)inTwo.readObject();  //得到原型的复制品
        }
        catch(Exception event){}
        return objectCopy;
    }
}
```

3）应用程序

下列应用程序（Application.java）复制了当前窗口中的容器。程序运行效果如图 25.4 所示。

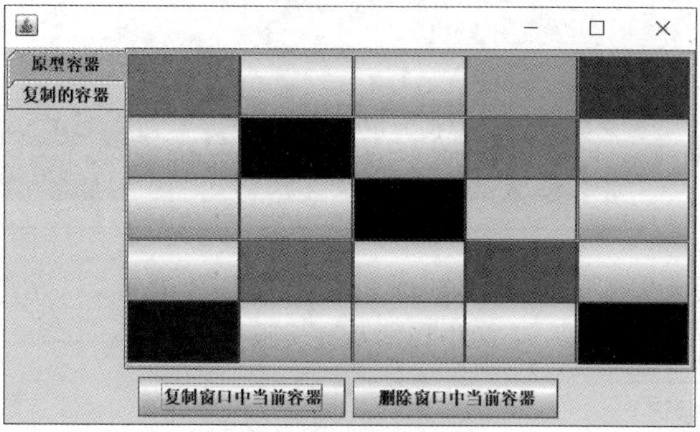

第25章 原型模式

图 25.4 程序运行效果

Application.java

```java
import javax.swing.*;
import java.awt.*;
public class Application extends JFrame {
    JTabbedPane jtp;
    CloneContainer con;                                          //原型
    JButton add,del;
    public Application(){
        add = new JButton("复制窗口中当前容器");
        del = new JButton("删除窗口中当前容器");
        add.addActionListener((e) ->{
            int index = jtp.getSelectedIndex();
            ButtonContainer container = (ButtonContainer)jtp.getComponentAt(index);
            CloneContainer conCopy = container.cloneContainer();//得到复制品
            jtp.add("复制的容器",(ButtonContainer)conCopy);
        });
        del.addActionListener((e) ->{
            int index = jtp.getSelectedIndex();
            ButtonContainer container = (ButtonContainer)jtp.getComponentAt(index);
            jtp.remove(index);
        });
        JPanel pSouth = new JPanel();
        pSouth.add(add);
        pSouth.add(del);
        add(pSouth,BorderLayout.SOUTH);
        con = new ButtonContainer();                             // 原型
        jtp = new JTabbedPane(JTabbedPane.LEFT);
        add(jtp,BorderLayout.CENTER);
        jtp.add("原型容器",(ButtonContainer)con);
        setBounds(100,100,500,300);
        setVisible(true);
        setDefaultCloseOperation(JFrame.EXIT_ON_CLOSE);
        validate();
    }
    public static void main(String args[]){
        new Application();
    }
}
```

第 26 章　备忘录模式

以下文本框中的内容引自 GoF 所著 *Design Patterns*：*Elements of Reusable Object Oriented Software* 的中译本及英文版。

> **备忘录模式**（别名：标记）
> 　　在不破坏封装性的前提下，捕获一个对象的内部状态，并在该对象之外保存这个状态，这样以后就可将该对象恢复到原先保存的状态。
> **Memento Pattern**（Another Name：Token）
> 　　Without violating encapsulation, capture and externalize an object'original state so that the object can be restored to this state later.

以上内容是 GoF 对备忘录模式的高度概括，结合 26.2.1 节的备忘录模式的类图可以准确地理解该模式。

26.1　概述

在玩游戏时，可能需要经过许多关卡才能最后成功，那么该游戏软件应当提供保存"游戏关卡"的功能，使游戏者在成功完成游戏的某一关卡之后，保存当前的游戏状态；当玩下一关卡失败时，可以选择让游戏从上一次保存的状态开始，即从上一次成功后的关卡开始，而不是再从第 1 关开始。游戏的状态与备忘录的示意图如图 26.1 所示。

图 26.1　游戏的状态与备忘录

备忘录模式是关于怎样保存对象状态的成熟模式，其关键是提供一个备忘录。该备忘录负责存储对象的状态，程序可以在磁盘或内存中保存这个备忘录。这样一来，程序就可以根据对象的备忘录将该对象恢复到备忘录中所存储的状态。

26.2　模式的结构与使用

备忘录模式包括三种角色。

1. 原发者（Originator）

原发者角色是一个类，该类的实例称为一个原发者，原发者要在某个时刻保存其状态。原发者负责创建备忘录,但不负责保存备忘录,备忘录由"负责人"统一管理。当原发者需要恢复某个时刻的状态时,"负责人"将提供备忘录,原发者通过备忘录恢复曾记录的状态。

2. 备忘录(Memento)

备忘录角色是一个类,该类的实例称为一个备忘录,备忘录负责存储原发者的状态。

3. 负责人(Caretaker)

负责人角色是一个类,该类的实例称为"负责人","负责人"负责管理备忘录,但不负责创建备忘录(负责人使用对象输入流、对象输出流管理备忘录)。

26.2.1 备忘录模式的 UML 类图

备忘录模式的类图如图 26.2 所示。

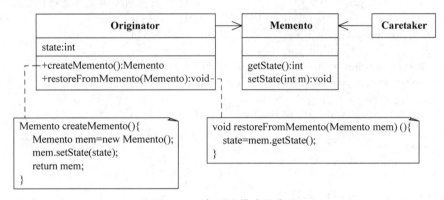

图 26.2 备忘录模式的类图

26.2.2 结构的描述

下面通过一个简单的问题来描述备忘录模式中所涉及的各个角色。

单词过关游戏:游戏每次出现一个单词,然后迅速消失,接着要求用户输入刚才出现的单词。如果输入正确,游戏进入下一级别,否则无法进入下一级别并提示是否保存目前的级别状态、退出游戏或继续尝试当前级别。下次重新开始游戏时,用户可以从保存的级别继续玩游戏。

1. 原发者

原发者是 Game 类。代码如下:

Game.java

```
import java.util.Scanner;
public class Game {
    int currentLevel = 0;                    //游戏的当前级别
    String word[] =                          //存放游戏中的单词
    {"boy","smile","river","great","quiet","reflection","caretaker"};

    public Memento createMemento(){
        Memento mem = new Memento();         //备忘录
        mem.setLevelState(currentLevel);     //备忘录存放当前级别状态
        return mem;
    }
    public void restoreFromMemento(Memento mem){   //从备忘录恢复状态
```

```java
            currentLevel = mem.getLevelState();
    }
    public boolean startGame(int index){
        int second = 5;
        currentLevel = index;
        System.out.printf
        ("这是第%d关,%d秒后,输入你看到的单词:\n",currentLevel + 1,second);
        System.out.print(word[currentLevel]);
        try {
            Thread.sleep(1000 * second);             //5000毫秒
        }
        catch(InterruptedException exp){}
        System.out.print("\r");                       //退行(回到行首,但不进行换行)
        for(int i = 0 ;i < word[currentLevel].length();i++){
            System.out.print("△");                    //消除单词
        }
        Scanner scanner = new Scanner(System.in);
        System.out.print("输入看到的单词(回车确认):");
        String strWord = scanner.nextLine();
        if(strWord.equals(word[currentLevel]))
            return true;
        else
            return false;
    }
    public int getHighestLevel(){
        return word.length - 1;
    }
}
```

2. 备忘录

备忘录是 Memento 类。代码如下:

Memento.java

```java
public class Memento implements java.io.Serializable{
    private int levelState;
    public void setLevelState(int level){
        if(level >= 0)
            levelState = level;
    }
    public int getLevelState(){
        return levelState;
    }
}
```

3. 负责人

负责人是 Caretaker 类。代码如下:

Caretaker.java

```java
import java.io.*;
public class Caretaker{
    File file;                                        //用于存放备忘录的文件
```

第26章 备忘录模式

```
        private Memento memento = null;
        Caretaker(){
            file = new File("saveObject.txt");
        }
        public Memento getMemento(){
            if(file.exists()) {
                try{
                    FileInputStream in = new FileInputStream("saveObject.txt");
                    ObjectInputStream inObject = new ObjectInputStream(in);
                    memento = (Memento)inObject.readObject();        //读取备忘录
                    inObject.close();
                    in.close();
                }
                catch(Exception exp){}
            }
            return memento;
        }
        public void saveMemento(Memento memento){
            try{
                FileOutputStream out = new FileOutputStream("saveObject.txt");
                ObjectOutputStream outObject = new ObjectOutputStream(out);
                outObject.writeObject(memento);                       //保存备忘录
                outObject.close();
                out.close();
            }
            catch(Exception exp){}
        }
        public void deleteFile(){
            file.delete();
        }
    }
```

▶ 26.2.3 模式的使用

前面已经使用备忘录模式给出了可以使用的类,这些类就是一个小框架,可以使用这个小框架中的类编写应用程序。

在应用程序 Application.java 中,程序每次出现一个单词,5 秒钟后消失,接着要求用户输入刚才出现的单词。如果输入正确,进入下一级别,否则无法进入下一级别并提示是否保存目前的级别状态,然后退出游戏。下次重新开始游戏时,用户可以从保存的级别开始继续玩游戏。程序运行效果如图 26.3 所示。

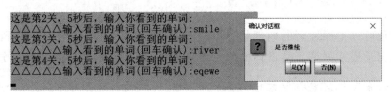

图 26.3 程序运行效果

Application.java

```java
import javax.swing.JOptionPane;
public class Application {
    public static void main(String args[]) {
        Game game = new Game();
        Caretaker caretaker = new Caretaker();
        int currentLevel = 0;
        int highestLevel = game.getHighestLevel();
        Memento memento = caretaker.getMemento();
        if(memento == null) {                                    //首次启动游戏
            currentLevel = 0;
        }
        else {
            currentLevel = memento.getLevelState();
        }
        boolean isSuccess = game.startGame(currentLevel);
        while(true) {
            if(isSuccess == false) {                             //在 currentLevel 级别失败
                int n = JOptionPane.showConfirmDialog(null,"是否继续","确认对话框",
                                    JOptionPane.YES_NO_OPTION );
                if(n == JOptionPane.YES_OPTION) {
                    isSuccess = game.startGame(currentLevel);    //保持在当前关
                }
                else if(n == JOptionPane.NO_OPTION){
                    caretaker.saveMemento(game.createMemento()); //保存备忘录
                    System.exit(0);
                }
            }
            else {
                currentLevel = currentLevel + 1;
                if(currentLevel > highestLevel){
                    System.out.println("成功通关!");
                    caretaker.deleteFile();                      //删除存储的备忘录
                    System.exit(0);
                }
                isSuccess = game.startGame(currentLevel);        //进入下一关
            }
        }
    }
}
```

26.3 备忘录模式的优点

使用备忘录可以把原发者的内部状态保存起来,使只有"亲密的"的对象可以访问备忘录中的数据。备忘录模式强调了类设计的单一责任原则,即将状态的刻画和保存分开。

注意:如果备忘录需要存储大量的数据或非常频繁地创建备忘录,可能会导致非常大的存储开销。

第26章 备忘录模式

扫一扫

视频讲解

26.4 应用举例——使用备忘录实现 undo 操作

1. 设计要求

使用备忘录模式来设计一个 GUI 程序,主要功能要求如下:

(1) 程序的窗体中有一个标签组件,用户在标签上单击鼠标可以在标签上随机显示一个汉字,但标签上只保留最后一次单击鼠标所显示的汉字。

(2) 程序提供 undo 操作。当用户在标签上右击鼠标时,将取消用户最近一次单击鼠标所产生的操作效果,即将标签上的汉字恢复为上一次单击鼠标所得到的汉字。用户可以多次右击鼠标依次取消单击鼠标所产生的操作效果。

2. 设计实现

1) 原发者与备忘录

原发者角色是 UnicodeLabel 类,UnicodeLabel 类是 javax.swing 包中 JLabel 类的子类,包含 Integer 对象,该对象中的 int 值代表一个汉字在 Unicode 表中的位置。

备忘录是 UnicodeLabel 的内部类。UnicodeLabel 类的代码如下:

UnicodeLabel.java

```java
import javax.swing.*;
import java.awt.*;
import java.awt.event.*;
public class UnicodeLabel extends JLabel{
    private int m;
    public UnicodeLabel(){
        setFont(new Font("宋体",Font.BOLD,72));
        setHorizontalAlignment(SwingConstants.CENTER);
        m = 19968;
        setText("" + (char)m);
        addMouseListener(new MouseAdapter(){
            public void mouseReleased(MouseEvent e) {
                if(e.getButton() == MouseEvent.BUTTON1){       //释放鼠标左键
                    m = (int)(Math.random() * 1000 + 19968);
                    setText("" + (char)m);
                }
            }});
    }
    public Memento createMemento(){
        Memento mem = new Memento();
        mem.setState(m);
        return mem;
    }
    public void restoreFromMemento(Memento mem){
        int n = mem.getState();
        setText("" + (char)n);
    }
    class Memento {                                            //Memento 是 UnicodeLabel 中的内部类
        int m;
        void setState(int m){
```

```
            this.m = m;
        }
        int getState(){
            return m;
        }
    }
}
```

2) 负责人

负责人是 Caretaker 类。Caretaker 类使用一个栈来存放备忘录,当用户需要 undo 操作时,从栈弹出最近一次的备忘录给用户,用户用该备忘录恢复原发者的状态;当栈为空时,用户不能进行 undo 操作。代码如下:

Caretaker.java

```java
import java.util.Stack;
public class Caretaker {
    Stack <UnicodeLabel.Memento> stack;
    Caretaker(){
        stack = new Stack <UnicodeLabel.Memento>();
    }
    public UnicodeLabel.Memento getMemento(){
        if(!(stack.isEmpty())){
            UnicodeLabel.Memento memento = stack.pop();
            return memento;
        }
        else{
            return null;
        }
    }
    public void saveMemento(UnicodeLabel.Memento memento){
        stack.push(memento);
    }
}
```

3) 应用程序

下列应用程序(Application.java)将原发者创建的标签添加到窗体中,当用户在标签上单击鼠标时,在标签上显示一个汉字,但标签上只保留最后一次单击鼠标所显示的汉字;当用户在标签上右击鼠标时,取消用户最近一次单击鼠标所产生的操作效果;当用户多次右击鼠标时,依次取消单击鼠标所产生的操作效果。程序运行效果如图 26.4 所示。

图 26.4　程序运行效果

Application. java

```java
import javax.swing.*;
import java.awt.*;
import java.awt.event.*;
public class Application extends JFrame implements MouseListener{
    UnicodeLabel label;                                              //原发者
    Caretaker caretaker;                                             //负责人
    Application(){
        label = new UnicodeLabel();
        label.addMouseListener(this);
        add(new JLabel("单击鼠标显示一个汉字,右击鼠标撤销单击鼠标的操作效果"),
                    BorderLayout.NORTH);
        add(label,BorderLayout.CENTER);
        caretaker = new Caretaker();                                 //创建负责人
    }
    public void mousePressed(MouseEvent e) {
        if(e.getButton() == MouseEvent.BUTTON1){                     //按下鼠标左键
            caretaker.saveMemento(label.createMemento());            //保存备忘录
        }
        if(e.getButton() == MouseEvent.BUTTON3){                     // 按下鼠标右键
            UnicodeLabel.Memento memento = caretaker.getMemento();   //得到备忘录
            if(memento!= null){
                label.restoreFromMemento(memento);                   //使用备忘录恢复状态
            }
        }
    }
    public void mouseReleased(MouseEvent e){}
    public void mouseEntered(MouseEvent e){}
    public void mouseExited(MouseEvent e){}
    public void mouseClicked(MouseEvent e){}
    public static void main(String args[]) {
        Application win = new Application();
        win.setBounds(10,10,300,300);
        win.setVisible(true);
        win.setDefaultCloseOperation(JFrame.EXIT_ON_CLOSE);
    }
}
```

第 27 章 享元模式

以下文本框中的内容引自 GoF 所著 *Design Patterns*：*Elements of Reusable Object Oriented Software* 的中译本及英文版。

> **享元模式**
> 运用共享技术有效地支持大量细粒度的对象。
> **Flyweight Pattern**
> Use sharing to support large numbers of fine grained objects efficiently.

以上内容是 GoF 对享元模式的高度概括，结合 27.2.1 节的享元模式的类图可以准确地理解该模式。

27.1 概述

一个类的成员变量表明该类所创建对象具有的属性。在某些程序设计中用一个类创建若干个对象，但是发现这些对象的一个共同特点是它们有一部分属性的取值必须是完全相同的。例如，一个 Car 类：

```
public class Car {
    double height,                //高度
           width,                 //宽度
           length;                //长度
    String color;                 //颜色
    double power;                 //功率
}
```

当用 Car 类创建若干个同型号的轿车时，例如创建若干个"奥迪 A6"轿车，要求这些轿车的 height、width、length 值必须相同（轿车的属性很多，属于细粒度对象，而且不同轿车的很多属性值是相同的，这里只示意了 height、width 和 length 三个属性），而 color、power 可以是不同的，即它们的长度、高度和宽度是相同的，但颜色和功率可能不同，如图 27.1 所示。

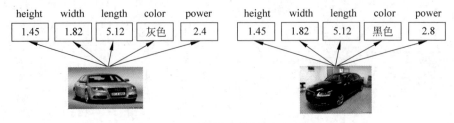

图 27.1 Car 类创建的"奥迪 A6"轿车

从创建对象的角度看,我们面对的问题是:Car 的每个对象的变量(height、width、length、color、power)都各自占有着不同的内存空间,这样一来,Car 创建的对象越多就越浪费内存空间,而且程序无法保证 Car 类创建的多个对象所对应的 height、width 和 length 值是相同的,或者禁止 Car 类创建的对象随意更改自己的 height、width、length 值。

享元模式的关键是使用一个享元为一些对象提供共享的状态,并且保证使用享元的这些对象不能更改享元中的数据。一个对象如果使用了享元,从享元的角度看,享元所维护的数据称作对象的内部状态,而对象的其他成员数据称作对象的外部状态,外部状态往往具有不可预见性,需要动态地计算确定。

27.2 模式的结构与使用

扫一扫

视频讲解

享元模式包括三种角色。

1. 享元(Flyweight)

享元是一个接口,定义了享元对外公开其内部数据的方法,以及享元接收外部数据的方法。

2. 具体享元(Concrete Flyweight)

具体享元角色是实现享元接口的类,该类的对象称为享元对象,简称享元。具体享元类的构造方法必须是 private 的,其目的是不允许用户程序直接使用具体享元类来创建享元,创建和管理享元由享元工厂负责。

3. 享元工厂(Flyweight Factory)

享元工厂是一个类,该类的实例负责创建和管理享元,用户或其他对象必须请求享元工厂为它分配一个享元。享元工厂通过散列表(也称作共享池)管理享元,当用户程序或其他若干个对象向享元工厂请求一个享元时,如果享元工厂的散列表中已有这样的享元,享元工厂就向请求者提供这个享元,否则就创建一个享元并添加到散列表中,同时将该享元提供给请求者。显然,当若干个用户或对象请求享元工厂提供一个享元时,第一个用户获得该享元的时间可能慢一些,但是后继的用户会较快地获得这个享元。可以使用单列模式设计享元工厂,即让系统中只有一个享元工厂的实例。另外,为了让享元工厂能生成享元,需要将具体享元类作为享元工厂的内部类。

> **注意**:如果一个对象中有享元,那么相对于享元(内部数据),称这个对象的其他成员变量是对象的外部数据。内部数据是不可以再发生改变的,外部数据可以发生变化。

▶ 27.2.1 享元模式的 UML 类图

享元模式的类图如图 27.2 所示。

▶ 27.2.2 结构的描述

下面通过一个简单的问题描述享元模式中所涉及的各个角色。

创建若干个"奥迪 A6"轿车和若干个"奥迪 A4"轿车。"奥迪 A6"轿车的长、宽和高都是相同的,颜色和功率可以不同;"奥迪 A4"轿车的长、宽和高都是相同的,颜色和功率可以不同。

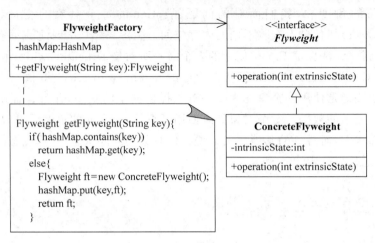

图 27.2　享元模式的类图

1. 享元

享元是 Flyweight 接口，定义了返回内部数据的方法和获得外部数据的方法。代码如下：

Flyweight.java

```
public interface Flyweight{
    public double getHeight();        //返回内部数据
    public double getWidth();         //返回内部数据
    public double getLength();        //返回内部数据
}
```

2. 享元工厂与具体享元

享元工厂是 FlyweightFactory 类，负责创建和管理享元对象。FlyweightFactory 将创建享元的具体享元类 ConcreteFlyweight 作为自己的内部类。FlyweightFactory 类的代码如下：

FlyweightFactory.java

```
import java.util.HashMap;
public class FlyweightFactory {
    private HashMap<String,Flyweight> hashMap;
    static FlyweightFactory factory = new FlyweightFactory();
    private FlyweightFactory(){
        hashMap = new HashMap<String,Flyweight>();
    }
    public static FlyweightFactory getFactory(){
        return factory;
    }
    public synchronized Flyweight getFlyweight(String key){
        if(hashMap.containsKey(key)){
            return hashMap.get(key);
        }
        else{
            double width = 0, height = 0, length = 0;
            String str[] = key.split("#");
            width = Double.parseDouble(str[0]);
```

```java
            height = Double.parseDouble(str[1]);
            length = Double.parseDouble(str[2]);
            Flyweight ft = new ConcreteFlyweight(width,height,length);
            hashMap.put(key,ft);
            return ft;
        }
    }
    class ConcreteFlyweight implements Flyweight{         //具体享元角色是内部类
        private double width;
        private double height;
        private double length;
        private ConcreteFlyweight(double width,double height,double length){
            this.width = width;
            this.height = height;
            this.length = length;
        }
        public double getHeight(){
            return height;
        }
        public double getWidth(){
            return width;
        }
        public double getLength(){
            return length;
        }
    }
}
```

▶ 27.2.3　模式的使用

前面已经使用享元模式给出了可以使用的类,这些类就是一个小框架,可以使用这个小框架中的类编写应用程序。

下列应用程序中,Car.java 使用 Car 类组合了享元(使用 Flyweight 类型的成员变量);Application.java 使用 Car 类分别创建了两辆奥迪 A6 轿车和两辆奥迪 A4 轿车。程序运行效果如图 27.3 所示。

```
名称:奥迪A6 颜色:黑色 功率:128 宽度:1.82 高度:1.47 长度:5.12
名称:奥迪A6 颜色:灰色 功率:160 宽度:1.82 高度:1.47 长度:5.12
名称:奥迪A4 颜色:蓝色 功率:126 宽度:1.77 高度:1.45 长度:4.63
名称:奥迪A4 颜色:红色 功率:138 宽度:1.77 高度:1.45 长度:4.63
```

图 27.3　程序运行效果

Car.java

```java
public class Car{
    Flyweight flyweight;                        //享元
    String name,color;
    int power;
    Car(Flyweight flyweight,String name,String color,int power){
        this.flyweight = flyweight;            //享元(内部数据)
```

```
            this.name = name;                                   //外部数据
            this.color = color;                                 //外部数据
            this.power = power;                                 //外部数据
        }
        public void print() {                                   //输出享元内部数据和外部数据
            System.out.print(" 名称:" + name);                  //输出外部数据(相对于享元)
            System.out.print(" 颜色:" + color);                 //输出外部数据
            System.out.print(" 功率:" + power);                 //输出外部数据
            System.out.print(" 宽度:" + flyweight.getWidth());  //输出享元内部数据
            System.out.print(" 高度:" + flyweight.getHeight()); //输出享元内部数据
            System.out.println("长度:" + flyweight.getLength());//输出享元内部数据
        }
    }
```

Application.java

```
public class Application{
    public static void main(String args[]) {
        FlyweightFactory factory = FlyweightFactory.getFactory();
        double width = 1.82, height = 1.47, length = 5.12;
        String key = "" + width + "#" + height + "#" + length;
        Flyweight flyweight = factory.getFlyweight(key);        //享元工厂提供一个享元
        Car audiA6One = new Car(flyweight,"奥迪 A6","黑色",128); //A6 使用同样的享元
        Car audiA6Two = new Car(flyweight,"奥迪 A6","灰色",160); //A6 使用同样的享元
        // audiA6One 和 audiA6Two 没有向享元传递外部数据,而是获取享元的内部数据
        audiA6One.print();
        audiA6Two.print();
        width = 1.77;
        height = 1.45;
        length = 4.63;
        key = "" + width + "#" + height + "#" + length;
        flyweight = factory.getFlyweight(key);                  //享元工厂又提供一个享元
        Car audiA4One = new Car(flyweight,"奥迪 A4","蓝色",126);
                                                                //A4 使用同样的享元(和 A6 不同)
        Car audiA4Two = new Car(flyweight,"奥迪 A4","红色",138);
                                                                //A4 使用同样的享元(和 A6 不同)
        audiA4One.print();
        audiA4Two.print();
    }
}
```

27.3 享元模式的优点

使用享元可以节省内存的开销,特别适合处理大量细粒度对象,这些对象的许多属性值是相同的,而且一旦创建不允许修改。

27.4 应用举例——创建化合物

1. 设计要求

氢氧化合物都是由氢元素和氧元素构成的,只是含有元素的个数不同。有人设计了如下

用于表示两种元素构成的化合物的 Compound 类：

```
public class Compound {
    char elementOne,                    //元素的名称
         elementTwo;                    //元素的名称
    int elementOneNumber,               //元素的个数
        elementTwoNumber;               //元素的个数
}
```

上述 Compound 类的设计有不合理之处，分析如下：

由于 Compound 类在创建若干个对象时，要求创建每个对象的 elementOne 变量和 elementTwo 变量的取值相同，例如创建若干个表示氢氧化合物的对象时，每个对象的 elementOne 变量和 elementTwo 变量的取值分别是 H 和 O，因此，为 Compound 类的每个对象的 elementOne 变量和 elementTwo 变量分配不同的内存空间是不必要的。

请使用享元模式重新设计 Compound 类。

2．设计实现

1) 享元

享元是 Element 接口。代码如下：

Element.java

```
public interface Element {
    public char getElementOne();        //返回内部数据
    public char getElementTwo();        //返回内部数据
}
```

2) 享元工厂与具体享元

享元工厂是 ElementFactory 类，负责创建和管理享元对象。ElementFactory 将创建享元对象的具体享元类 TwoElement 作为自己的内部类。ElementFactory 类的代码如下：

ElementFactory.java

```
import java.util.HashMap;
public class ElementFactory {
    private HashMap<String,Element> hashMap;
    static ElementFactory factory = new ElementFactory();
    private ElementFactory(){
        hashMap = new HashMap<String,Element>();
    }
    public static ElementFactory getFactory(){
        return factory;
    }
    public synchronized Element getElement(String key){
        if(hashMap.containsKey(key)) {
            return hashMap.get(key);
        }
        else{
            char elementOne = 0,elementTwo = 0;
            elementOne = key.charAt(0);
            elementTwo = key.charAt(1);
```

```
            Element element =
                new InnerElement(elementOne,elementTwo);            //具体享元是内部类
            hashMap.put(key,element);
            return element;
        }
    }
    class InnerElement implements Element{                          // InnerElement 是内部类
        char elementOne,elementTwo;
        private InnerElement(char elementOne,char elementTwo){
            this.elementOne = elementOne;
            this.elementTwo = elementTwo;
        }
        public char getElementOne(){
            return elementOne;
        }
        public char getElementTwo(){
            return elementTwo;
        }
    }
}
```

3) 应用程序

下列应用程序中，Compound.java 使用 Compound 类组合了享元（用 Element 成员作为自己的成员变量）；Application.java 使用 Compound 类分别创建了由元素碳（C）和元素氧（O）构成的"二氧化碳"和"一氧化碳"，以及由元素氢（H）和元素氧（O）构成的"水"和"过氧化氢"。程序运行效果如图 27.4 所示。

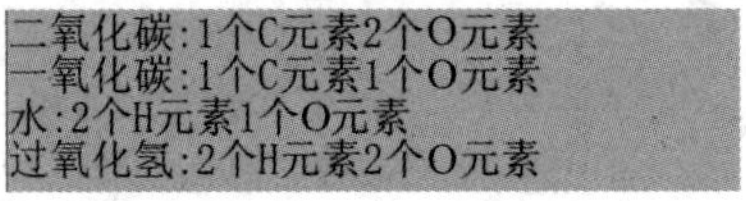

```
二氧化碳:1个C元素2个O元素
一氧化碳:1个C元素1个O元素
水:2个H元素1个O元素
过氧化氢:2个H元素2个O元素
```

图 27.4　程序运行效果

Compound.java

```java
public class Compound{
    Element element;                                    //存放享元对象的引用
    String name;                                        // 相对于享元,是对象的外部数据
    int number1,number2;                                // 相对于享元,是对象的外部数据
    Compound(Element element,String name,int number1,int number2){
        this.element = element;                         //享元
        this.name = name;                               //外部数据
        this.number1 = number1;                         //外部数据
        this.number2 = number2;                         //外部数据
    }
    public void printElement() {                        //输出内部数据(享元)和外部数据
        System.out.print(name + ":" + number1 + "个" +
                        element.getElementOne() + "元素");
        System.out.println(number2 + "个" +
                        element.getElementTwo() + "元素");
    }
}
```

Application.java

```java
public class Application{
    public static void main(String args[]) {
        ElementFactory factory = ElementFactory.getFactory();
        String key = "CO",name;
        Element element = factory.getElement(key);        //提供一个享元(C 和 O 元素)
        Compound compound = new Compound(element,"二氧化碳",1,2);
        compound.printElement();
        compound = new Compound(element,"一氧化碳",1,1);
        compound.printElement();
        key = "HO";
        element = factory.getElement(key);                //提供一个享元(H 和 O 元素)
        compound = new Compound(element,"水",2,1);
        compound.printElement();
        compound = new Compound(element,"过氧化氢",2,2);
        compound.printElement();
    }
}
```

第 28 章 MVC 模式

MVC 模式不是 GoF 所著 *Design Patterns: Elements of Reusable Object Oriented Software* 一书中的 23 个模式之一,但已经有很多成功的案例。

28.1 概述

Model-View-Controller(模型-视图-控制器,MVC)是 Xerox PARC 在 20 世纪 80 年代为编程语言 Smalltalk-80 发明的一种软件编程模式,现今已经成为软件开发者必须熟练使用的开发模式。例如,在 Java Server Page(JSP)技术中,就要熟悉 MVC 模式。

从面向对象的角度看,MVC 模式使程序容易维护,也更容易扩展。在编写程序时,可以将某个对象看作"模型",然后为"模型"提供恰当的显示组件,即"视图"。在 MVC 模式中,"视图""模型"和"控制器"之间是松耦合结构,便于系统的维护和扩展。

本章主要以 Java 基础知识为主(JDK),学习使用 MVC 模式,体会 MVC 模式的编程思想。

扫一扫

视频讲解

28.2 模式的结构与使用

MVC 包括三种角色。

1. 模型(model)

模型用于存储数据的若干个对象。

2. 视图(View)

视图由若干个可视组件构成。视图负责向控制器提交所需数据、显示模型中的数据。

3. 控制器(controller)

控制器负责处理数据的若干个对象。这些对象负责具体的业务逻辑操作,即控制器根据视图提出的要求对数据做出(商业)处理,将有关结果存储到模型中,并负责让模型和视图进行必要的交互(当模型中的数据变化时,让视图更新显示)。

▶ 28.2.1 MVC 模式的示意图

严格来讲,MVC 不属于设计模式,而是一种编程思想或思路。因此,MVC 和前面章节的设计模式不同,MVC 没有严格的类的 UML 图。MVC 的应用范围也很广泛,其中的各个角色也不一定是类或接口。所以,对于 MVC 模式,只能给出 MVC 的示意图,即根据 MVC 模式各个角色的职责绘制的简单示意图。MVC 模式的示意图如图 28.1 所示。

第28章 MVC模式

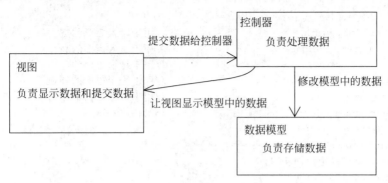

图 28.1 MVC 模式的示意图

28.2.2 结构的描述

下面通过求矩形的面积描述 MVC 模式中所涉及的各个角色。

1. 数据模型

模型是 Rectangle 类，提供了矩形相关的方法。代码如下：

Rectangle.java

```java
public class Rectangle {
    public double width,             //矩形的宽
                  height;            //矩形的高
    public void setHeight(double h) {
        if(h > 0)
            height = h;
    }
    public double getHeight() {
        return height;
    }
    public void setWidth(double w) {
        if(w > 0)
            width = w;
    }
    public double getWidth() {
        return width;
    }
    public double getArea() {
        return width * height;
    }
}
```

2. 视图

视图包括 RectangleView 类和 ShowShap 类。RectangleView 类的实例（需要组合控制器）负责提交数据给控制器，用文本显示模型中的数据；ShowShape 类的实例用图形显示模型中的数据。代码分别如下：

RectangleView.java

```java
import javax.swing.*;
import java.awt.*;
```

```java
public class RectangleView extends JFrame {
    JTextField inputWidth,                      //用于输入矩形的宽
              intputHeight,                     //用于输入矩形的高
              showArea;                         //用于显示矩形的面积
    ShowShape showShape;                        //用于绘制矩形的外观
    JButton submit;                             //用于提交数据给控制器
    Controller controller;                      //组合控制器
    public RectangleView() {
        initComponent();
        setTitle("求矩形面积");
        setBounds(10,10,700,350);
        setVisible(true);
        setDefaultCloseOperation(JFrame.EXIT_ON_CLOSE );
    }
    public void initComponent(){
        inputWidth = new JTextField(10);
        intputHeight = new JTextField(10);
        showArea = new JTextField(10);
        showArea.setEditable(false);
        showShape = new ShowShape();
        submit = new JButton("提交数据给控制器");
        controller = new Controller();
        controller.setView(this);
        JPanel pNorth = new JPanel();
        pNorth.add(new JLabel("矩形的宽:"));
        pNorth.add(inputWidth);
        pNorth.add(new JLabel("矩形的高:"));
        pNorth.add(intputHeight);
        pNorth.add(submit);
        pNorth.add(new JLabel("矩形的面积:"));
        pNorth.add(showArea);
        add(pNorth,BorderLayout.NORTH);
        add(showShape,BorderLayout.CENTER);
        submit.addActionListener(controller);    //单击按钮,提交数据给控制器
    }
}
```

ShowShape.java

```java
import java.awt.*;
import javax.swing.*;
public class ShowShape extends JPanel {
    double width,
           height;
    public void paint(Graphics g ) {
        super.paint(g);
        g.drawRect(10,10,(int)width*5,(int)height*5);
    }
    public void setWidth(double w) {
        if(w > 0)
            width = w;
```

```java
    public void setHeight(double h) {
        if(h > 0)
            height = h;
    }
}
```

3. 控制器

控制器是 Controller 类,该类的实例负责处理视图提交给它的数据和数据模型中的数据,然后让视图显示模型中的数据。代码如下:

Controller.java

```java
import java.awt.event.ActionEvent;
import java.awt.event.ActionListener;
public class Controller implements ActionListener {
    public RectangleView view ;                    //组合视图
    public Rectangle tangle ;                      //组合数据模型
    public void setView(RectangleView view) {
        this.view = view;
    }
    public void actionPerformed(ActionEvent exp) {  //获得视图提交的数据
        double width = 0;
        double height = 0;
        double area = 0 ;
        try {
            width = Double.parseDouble(view.inputWidth.getText());
            height = Double.parseDouble(view.intputHeight.getText());
            tangle = new Rectangle();              //处理数据模型
            tangle.setWidth(width);
            tangle.setHeight(height);
            area = tangle.getArea();
            view.showArea.setText("" + area);      //让视图更新数据的显示
            view.showShape.setWidth(width);
            view.showShape.setHeight(height);
            view.showShape.repaint();
        }
        catch(NumberFormatException e) {
            view.showArea.setText("请输入数字字符");
        }
    }
}
```

▶ 28.2.3 模式的使用

前面已经使用 MVC 模式给出了可以使用的类,这些类就是一个小框架,可以使用这个小框架中的类编写应用程序。

下列应用程序(Application.java)使用 RectangleView 类创建了一个窗口。程序运行效果如图 28.2 所示。

图 28.2 程序运行效果

Application.java

```
public class Application{
    public static void main(String args[]) {
        RectangleView view = new RectangleView();
    }
}
```

28.3 MVC 模式的优点

MVC 模式的应用范围相当广泛，尤其在 Web 领域，例如在 JSP 中，数据模型是 bean（类似 28.2.2 节中的 Rectangle），视图是 JSP 页面（类似 28.2.2 节中的 RectangleView），控制器是 servlet（类似 28.2.2 节中的 Controller）。MVC 不仅可以使用各种不同样式的视图共享一个数据模型，以适应用户的需求，而且使程序容易维护，也更容易扩展。

28.4 应用举例——老鼠走迷宫

1. 设计要求

在 7.4 节的应用举例中，使用策略模式设计了老鼠走迷宫，但是当时的程序只能在命令行显示老鼠走迷宫的过程。请使用 MVC 模式提供更好的视图，演示老鼠走迷宫的过程。

2. 设计实现

1) 数据模型

数据模型是 7.4 节中的类，但做了非常微小的改动，以便适合视图显示模型中的数据。数据模型包括 MazeStrategy 类、GreedyStrategy 类和 StackStrategy 类。代码分别如下：

MazeStrategy.java

```
import java.awt.Point;
import java.util.ArrayList;
public interface MazeStrategy{
    //相对7.4节代码增加一行代码,存放老鼠走过的路点
    public static ArrayList<Point> stepPoint = new ArrayList<Point>();
```

```java
    public static char ROAD = '路';
    public static char WALL = '■';
    public static char EXIT = '出';
    public abstract char [][] moveInMaze(char maze[][]);          //走迷宫的方法
    public default char [][] copyArray(char a[][]) {
        char maze[][] = new char[a.length][a[0].length];
        for(int i = 0;i < a.length;i++) {
            for(int j = 0;j < a[0].length;j++)
                maze[i][j] = a[i][j];
        }
        return maze;
    }
}
```

GreedyStrategy.java

```java
import java.awt.Point;                                            //相对 7.4 节代码增加一行代码
public class GreedyStrategy implements MazeStrategy{
    public char [][] moveInMaze(char a[][]){
        char mazeChar[][] = copyArray(a);
        int Y = Integer.MAX_VALUE;
        int N = Integer.MIN_VALUE;
        int mazeInt[][] = changeToInt(mazeChar,Y,N);
        int mousePI = 0;                                          //老鼠初始位置是左上角
        int mousePJ = 0;
        int rows;
        int columns;
        rows = mazeChar.length;
        columns = mazeChar[0].length;
        stepPoint.clear();
        while(mazeInt[mousePI][mousePJ]!= Y){
            stepPoint.add(new Point(mousePI,mousePJ));            //相对 7.4 节代码增加一行代码
            mazeChar[mousePI][mousePJ] = '走';
            mazeInt[mousePI][mousePJ] -- ;
            if(mazeInt[mousePI][mousePJ] == Y){
                System.out.println("陷入局部最优解");
                break;
            }
            int m = mousePI;
            int n = mousePJ;
            int max = mazeInt[mousePI][mousePJ];
            if(mousePJ > = 1){                                    //检查西方
                if(mazeInt[mousePI][mousePJ - 1]> max){
                    max = mazeInt[mousePI][mousePJ - 1];
                    m = mousePI;
                    n = mousePJ - 1;
                }
            }
            if(mousePI > = 1){                                    //北方
                if(mazeInt[mousePI - 1][mousePJ]> max){
                    max = mazeInt[mousePI - 1][mousePJ];
```

```java
                    m = mousePI - 1;
                    n = mousePJ;
                }
            }
            if(mousePJ < columns - 1){ //东方
                if(mazeInt[mousePI][mousePJ + 1] > max){
                    max = mazeInt[mousePI][mousePJ + 1];
                    m = mousePI;
                    n = mousePJ + 1;
                }
            }
            if(mousePI < rows - 1){ //南方
                if(mazeInt[mousePI + 1][mousePJ] > max){
                    max = mazeInt[mousePI + 1][mousePJ];
                    m = mousePI + 1;
                    n = mousePJ;
                }
            }
            mousePI = m;
            mousePJ = n;
        }
        mazeChar[mousePI][mousePJ] = '出';
        //相对7.4节代码增加一行代码:
        stepPoint.add(new Point(mousePI,mousePJ));
        return mazeChar;
    }
    private int [][] changeToInt(char[][] mazeChar, int Y, int N){
        int rows = mazeChar.length;
        int columns = mazeChar[0].length;
        int mazeInt[][] = new int[rows][columns];
        for(int i = 0;i < rows;i++){
            for(int j = 0;j < columns;j++) {
                if(mazeChar[i][j] == ROAD) {
                    mazeInt[i][j] = Y - 1;
                }
                else if(mazeChar[i][j] == WALL){
                    mazeInt[i][j] = N;
                }
                else if(mazeChar[i][j] == EXIT){
                    mazeInt[i][j] = Y;
                }
            }
        }
        return mazeInt;
    }
}
```

StackStrategy.java

```java
import java.util.Stack;
import java.awt.Point;
```

第28章 MVC模式

```java
public class StackStrategy implements MazeStrategy {
    public char [][] moveInMaze(char a[][]){
        boolean isSuccess = false;                    //是否走迷宫成功
        char maze[][] = copyArray(a);
        int rows = maze.length;
        int columns = maze[0].length;
        int x = 0;                                     //老鼠初始位置
        int y = 0;                                     //老鼠初始位置
        stepPoint.clear();
        Stack<Point> stack = new Stack<Point>();
        Point point = new Point(x,y);
        stack.push(point);                             //stack 进行压栈操作
        while(isSuccess == false) {                    //未走到迷宫出口
            if(!stack.empty())
                point = stack.pop();
            else {
                System.out.printf("无法到达出口,老鼠回到入口");
                return maze;
            }
            x = (int)point.getX();
            y = (int)point.getY();
            if(maze[x][y] == EXIT) {                   //是出口
                isSuccess = true;
                maze[x][y] = '出';
                stepPoint.add(new Point(x,y));         //相对7.4节代码增加一行代码
            }
            else {
                maze[x][y] = '走';
                stepPoint.add(new Point(x,y));         //相对7.4节代码增加一行代码
                if(y-1 >= 0&&(maze[x][y-1] == ROAD||maze[x][y-1] == EXIT)) {   //西是路
                    stack.push(new Point(x,y-1));
                }
                if(x-1 >= 0&&(maze[x-1][y] == ROAD||maze[x-1][y] == EXIT)) {   //北是路
                    stack.push(new Point(x-1,y));
                }
                if(y+1 < columns&&(maze[x][y+1] == ROAD||maze[x][y+1] == EXIT)){
                                                                                //东是路
                    stack.push(new Point(x,y+1));
                }
                if(x+1 < rows&&(maze[x+1][y] == ROAD||maze[x+1][y] == EXIT)){  //南是路
                    stack.push(new Point(x+1,y));     //stack 进行压栈操作
                }
            }
        }
        return maze;
    }
}
```

2) 视图

视图是 MazeView 类。代码如下:

MazeView.java

```java
import javax.swing.*;
import java.awt.*;
import java.util.ArrayList;
public class MazeView extends JFrame implements Runnable{
    ControllerFactory factory ;                    //控制器工厂
    Thread moveMouse;                              //负责移动老鼠外观的线程
    ArrayList<Point> list;
    char a[][];                                    //迷宫数据
    int w = 30;                                    //迷宫格的宽
    int h = 30;                                    //迷宫格的高
    int rows ;                                     //迷宫的行数
    int colums ;                                   //迷宫的列数
    int m;                                         //老鼠的当前位置的索引——行坐标
    int n;                                         //老鼠的当前位置的索引——列坐标
    int offX = 160,                                //绘图的偏移量
        offY = 100;
    ShowMaze showMaze;                             //内部类,负责显示走迷宫过程
    JMenuBar menubar;
    JMenu menu;
    JMenuItem simpleItem,complexItem;
    public MazeView() {
        initMenu();                                //初始化菜单
        moveMouse = new Thread(this);
        showMaze = new ShowMaze();
        add(showMaze,BorderLayout.CENTER);
        setBounds(10,10,800,500);
        setVisible(true);
        validate();
        setDefaultCloseOperation(JFrame.EXIT_ON_CLOSE );
    }
    public void initMenu(){
        menubar = new JMenuBar();
        menu = new JMenu("选择迷宫");
        simpleItem = new JMenuItem("简单迷宫");
        factory = new SimpeFactory();
        Controller controller = factory.getController();
        controller.setView(this);
        simpleItem.addActionListener(controller);  //提交数据给监视器
        complexItem = new JMenuItem("复杂迷宫");
        factory = new ComplexFactory();
        controller = factory.getController();
        controller.setView(this);
        complexItem.addActionListener(controller);
        menubar.add(menu);                         //菜单条添加菜单
        menu.add(simpleItem);
        menu.add(complexItem);
        setJMenuBar(menubar);                      //窗口放置菜单条
    }
    public void showMaze(char maze[][]){           //显示迷宫
        a = maze;
```

```java
            rows = a.length;
            colums = a[0].length;
            showMaze.repaint();
        }
        public void startMove(ArrayList<Point> list) {        //老鼠走迷宫的视图效果
            this.list = list;
            if(!moveMouse.isAlive()) {
                moveMouse = new Thread(this);
                moveMouse.start();
            }
        }
        public void run(){
            for(int i = 0;i < list.size();i++){
                m = (int)list.get(i).getX();
                n = (int)list.get(i).getY();
                System.out.printf("(%d,%d),",m,n);
                showMaze.repaint();
                try{
                    Thread.sleep(300);
                }
                catch(InterruptedException exp){}
            }
        }
        public class ShowMaze extends JPanel {                //内部类
            public void paint(Graphics g){
                super.paint(g);
                g.setColor(Color.red);
                g.fillOval(offX + n * w,offY + m * h,w,h);    //绘制老鼠外观(一个圆)
                for(int i = 0;i < rows;i++){
                    for(int j = 0;j < colums;j++) {
                        if(a[i][j] == MazeStrategy.WALL){
                            g.setColor(Color.black);
                            g.fillRect(offX + j * w,offY + i * h,w,h);
                        }
                        else if(a[i][j] == MazeStrategy.ROAD){
                            g.setColor(Color.green);
                            g.drawRect(offX + j * w,offY + i * h,w,h);
                        }
                        else if(a[i][j] == MazeStrategy.EXIT){
                            g.setColor(Color.blue);
                            g.fillRect(offX + j * w,offY + i * h,w,h);
                        }
                    }
                }
                g.setColor(Color.red);
                g.fillOval(offX + n * w,offY + m * h,w,h);
            }
        }
    }
}
```

3) 控制器

由于老鼠可以选择某种策略走某个迷宫，因此，这里采用工厂方法模式设置各种控制器（有关工厂模式的知识点见第13章）。Controller 类是抽象产品角色；ControllerSimple 类和 ControllerComplex 类是具体产品角色。ControllerFactory 类是构造者角色；SimpleFactory 类和 ComplexFactory 类是具体构造者角色（读者可以通过文件初始化迷宫或使用随机算法给出迷宫，这里为了减少代码量，使用了固定迷宫）。代码分别如下：

Controller.java

```java
import java.awt.event.ActionEvent;
import java.awt.event.ActionListener;
public abstract class Controller implements ActionListener {    //抽象产品
    public MazeView view ;                                       //组合视图
    public MazeStrategy strategy ;                               //组合数据模型
    public void setView(MazeView view) {
        this.view = view;
    }
}
```

ControllerSimple.java

```java
import java.awt.event.ActionEvent;
import java.awt.event.ActionListener;
import java.util.Arrays;
public class ControllerSimple extends Controller {              //具体产品
    public void actionPerformed(ActionEvent exp) {
        final char 路 = MazeStrategy.ROAD;
        final char 墙 = MazeStrategy.WALL;
        final char 出 = MazeStrategy.EXIT;
        char maze[][] = { { 路,路,墙,墙,墙,路 },
                          { 路,路,路,路,路,路 },
                          { 墙,路,墙,墙,路,路 },
                          { 路,路,墙,出,路,墙 }};
        strategy = new GreedyStrategy();                         //操作数据模型(使用贪婪策略算法)
        char a[][] = strategy.moveInMaze(maze);
        view.showMaze(maze);                                     //让视图更新
        view.startMove(strategy.stepPoint);                      //视图不需数组 a,需要 stepPoint
        //命令行显示需要数组 a(见 7.4 节)
        System.out.println("\"走\"和\"出\"表示老鼠曾走过的路和到达的出口:");
        for(int i = 0;i<a.length;i++){
            System.out.println(Arrays.toString(a[i]));
        }
    }
}
```

ControllerComplex.java

```java
import java.awt.event.ActionEvent;
import java.awt.event.ActionListener;
import java.util.Arrays;
public class ControllerComplex extends Controller {             //具体产品
```

```java
    public void actionPerformed(ActionEvent exp) {
        final char 路 = MazeStrategy.ROAD;
        final char 墙 = MazeStrategy.WALL;
        final char 出 = MazeStrategy.EXIT;
        char maze[][] = { { 路,路,墙,墙,墙,路,路,路,墙,墙,墙,墙 },
                          { 路,路,路,路,路,路,路,路,路,墙,路,墙 },
                          { 墙,路,墙,墙,路,路,墙,路,墙,路,墙 },
                          { 路,路,墙,墙,路,路,路,墙,墙,墙,墙 },
                          { 路,路,墙,墙,路,路,路,墙,路,墙,墙,路 },
                          { 路,路,墙,墙,路,路,路,墙,路,墙,墙 },
                          { 墙,路,墙,墙,路,路,路,墙,路,墙,路 },
                          { 墙,路,墙,墙,路,路,路,路,墙,出,墙 }};
        strategy = new StackStrategy();             //操作数据模型(使用栈策略算法)
        char a[][] = strategy.moveInMaze(maze);
        view.showMaze(maze);                        //让视图更新
        view.startMove(strategy.stepPoint);         //视图需要 stepPoint
        //命令行显示需要数组 a(见 7.4 节)
        System.out.println("\"走\"和\"出\"表示老鼠曾走过的路和到达的出口:");
        for(int i = 0;i < a.length;i++){
            System.out.println(Arrays.toString(a[i]));
        }
    }
}
```

ControllerFactory. java

```java
public interface ControllerFactory {            //构造者
    public abstract Controller getController(); //"工厂方法"返回一个控制器
}
```

SimpleFactory. java

```java
public class SimpleFactory implements ControllerFactory{    //具体构造者
    public Controller getController(){
        return new ControllerSimple();
    }
}
```

ComplexFactory. java

```java
public class ComplexFactory implements ControllerFactory{   //具体构造者
    public Controller getController(){
        return new ControllerComplex();
    }
}
```

4) 应用程序

前面用 MVC 模式给出了框架。下列应用程序(Application. java)使用框架中的 MazeView 创建了一个走迷宫的窗口视图。程序运行效果如图 28.3 所示。

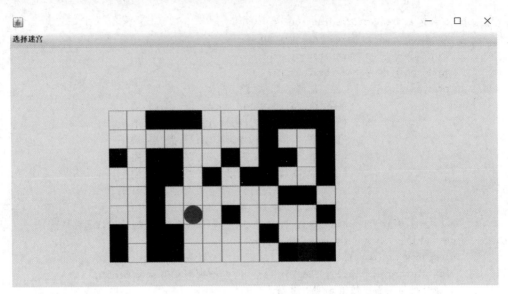

图 28.3　程序运行效果

Application. java

```
public class Application {
    public static void main(String args[]) {
        MazeView view = new MazeView();
    }
}
```

参 考 文 献

[1] Cooper J W. Java Design Patterns：A Tutorial[M]. Boston, MA：Addison Wesley, 2000.
[2] Gamma E, Helm R, Johnson R, et al. 设计模式：可复用面向对象软件的基础[M]. 李英军, 马晓星, 蔡敏, 等译. 北京：机械工业出版社, 2000.
[3] Metsker S J. 设计模式 Java 手册[M]. 龚波, 冯军, 程群梅, 等译. 北京：机械工业出版社, 2006.
[4] Cooper J W. Java 设计模式[M]. 王宇, 林琪, 杜志秀, 译. 北京：中国电力出版社, 2003.
[5] Kuchana P. Java 软件体系结构设计模式标准指南[M]. 王卫军, 楚宁志, 等译. 北京：电子工业出版社, 2006.
[6] Horstmann C. 面向对象的设计与模式[M]. 张琛恩, 译. 北京：电子工业出版社, 2004.
[7] Budd T A. 面向对象编程导论（原书第 3 版）[M]. 黄明军, 李桂杰, 译. 北京：机械工业出版社, 2003.

图书资源支持

感谢您一直以来对清华版图书的支持和爱护。为了配合本书的使用,本书提供配套的资源,有需求的读者请扫描下方的"书圈"微信公众号二维码,在图书专区下载,也可以拨打电话或发送电子邮件咨询。

如果您在使用本书的过程中遇到了什么问题,或者有相关图书出版计划,也请您发邮件告诉我们,以便我们更好地为您服务。

我们的联系方式:

地　　址:北京市海淀区双清路学研大厦 A 座 714

邮　　编:100084

电　　话:010-83470236　010-83470237

客服邮箱:2301891038@qq.com

QQ:2301891038(请写明您的单位和姓名)

资源下载:关注公众号"书圈"下载配套资源。

资源下载、样书申请

书　圈

图书案例

清华计算机学堂

观看课程直播